Practical Process Control for Engineers and Technicians

Dedication

This book is dedicated to Wolfgang who fought a courageous battle against motor neurone disease and continued teaching until the very end. Although he received his training in Europe, he ended up being one of Australia's most outstanding instructors in industrial process control and inspired IDC Technologies into running his course throughout the world. His delight in taking the most complex control system problems and reducing them to simple practical solutions made him a sought after instructor in the process control field and an outstanding mentor to the IDC Technologies engineers teaching the topic.

Hambani Kahle (Zulu Farewell)

(*Sources*: *Canciones de Nuestra Cabana* (1980), *Tent and Trail Songs* (American Camping Association), *Songs to Sing & Sing Again* by Shelley Gorden)

> Go well and safely.
> Go well and safely.
> Go well and safely.
> The Lord be ever with you.
>
> Stay well and safely.
> Stay well and safely.
> Stay well and safely.
> The Lord be ever with you.
>
> Hambani Kahle.
> Hambani Kahle.
> Hambani Kahle.
> The Lord be ever with you.

Practical Process Control for Engineers and Technicians

Wolfgang Altmann Dipl.Ing

Contributing author: David Macdonald BSc (Hons) Inst. Eng,
Senior Engineer, IDC Technologies, Cape Town, South Africa

Series editor: Steve Mackay FIE (Aust), CPEng, BSc (ElecEng), BSc (Hons),
MBA, Gov.Cert.Comp., Technical Director – IDC Technologies Pty Ltd

AMSTERDAM • BOSTON • HEIDELBERG • LONDON
NEW YORK • OXFORD • PARIS • SAN DIEGO
SAN FRANCISCO • SINGAPORE • SYDNEY • TOKYO

Newnes is an imprint of Elsevier

Newnes
An imprint of Elsevier
Linacre House, Jordan Hill, Oxford OX2 8DP
30 Corporate Drive, Burlington, MA 01803

First published 2005

British Library Cataloguing in Publication Data
A catalogue record for this book is available from the British Library

Library of Congress Cataloguing in Publication Data
A catalogue record for this book is available from the Library of Congress

ISBN 0 7506 6400 2

For information on all Newnes publications
visit our website at www.newnespress.com

Typeset by Integra Software Services Pvt. Ltd, Pondicherry, India
www.integra-india.com

Printed and bound in the United Kingdom

Transferred to Digital Print 2011

Contents

Contents x

Preface

Experience shows that most graduate engineers have a sound knowledge of the mathematical aspects of process control. Nevertheless, when it comes to the practical understanding of industrial process control, there is often a problem in converting this theoretical knowledge into a practical understanding of control concepts and problems. This publication, is intended to fill this gap.

It is not intended to add another book to the vast number of existing books, covering process control theory. Instead, this book provides a practical understanding of control concepts as well as enabling the reader to gain a correct understanding of control theory.

The principles of industrial process control concepts and the associated pitfalls are explained in an easy to understand manner. Although the mathematical side is kept to a minimum, a basic grasp of engineering concepts and a general knowledge of algebra and calculus is required in order to obtain a full understanding of this publication.

There is a degree of emphasis on the internal calculation of control algorithms in digital computers. The purpose of this is to provide a wider view of the use and modification of computer-calculated algorithms (incremental algorithms).

The first automatic control system known was the Fly-ball governor installed on Watt's steam engine in 1775 to regulate the steam rate. It was nearly a century later that the first mathematical model of the Fly-ball governor was prepared by James Clerk Maxwell. This illustrates a common practice in the development of process control, using a system before fully understanding exactly why and how it does the job. The spreading use of steam boilers resulted in the introduction of other automatic control systems, such as steam pressure regulators and the first multiple element boiler feedwater systems. Again, the applications came before the theory.

The first general theory of automatic control, written by Nyquist, only appeared in 1932.

Today, automatic control is an increasingly important part of the capital outlay in industry. The primary difficulty encountered with process control is in applying well-defined mathematical theories to day-to-day industrial applications, and translating ideal models to the frequently far-from ideal real-world scenario.

Process control has a number of significant advantages. As always, the primary factor in any operation is cost. The use of process control in a system enables the maximum profitability to be derived. Other advantages are that automatic control results in increased plant flexibility, reduced maintenance, and in stable and safe operation of the plant.

It also allows operators to more closely approach optimum operation of the process. As the degree of automatic control is increased, so do the related advantages which too become more significant.

Further improvements in process control are attained by model-based control and ultimately by optimization.

Optimization applications can be installed when the plant is stable, operated safely and has tight quality control. The benefits of optimization are improvement in product yield and quality, reduction in energy consumption, and a move to optimum operation of the process. It is possible to track optimum operation to maintain the maximum profitability of the process.

The Industrial Process Control Software used in this course (Windows or DOS) is available by either going to the IDC Technologies web site located at: www.idc-online.com

or by emailing IDC Technologies at: idc@idc-online.com

1

Introduction

1.1 Objectives

As a result of studying this chapter, the student should be able to:

- Describe the three different types of processes
- Indicate the meaning of a time constant
- Describe the meaning of process variable, setpoint and output
- Outline the meaning of first and second order systems
- List the different modes of operation of a control system.

1.2 Introduction

To succeed in process control the designer must first establish a good understanding of the process to be controlled. Since we do not wish to become too deeply involved in chemical or process engineering, we need to find a way of simplifying the representation of the process we wish to control. This is done by adopting a technique of block diagram modeling of the process.

All processes have some basic characteristics in common, and if we can identify these, the job of designing a suitable controller can be made to follow a well-proven and consistent path. The trick is to learn how to make a reasonably accurate mathematical model of the process and use this model to find out what typical control actions we can use to make the process operate at the desired conditions.

Let us then start by examining the component parts of the more important dynamics that are common to many processes. This will be the topic covered in the next few sections of this chapter, and upon completion we should be able to draw a block diagram model for a simple process; for example, one that says: 'It is a system with high gain and a first order dynamic lag and, as such, we can expect it to perform in the following way', regardless of what the process is manufacturing or its final product.

From this analytical result, an accurate selection of the type of measuring transducer can be selected, this being covered in Chapter 2. Likewise, the selection of the final control element can be correctly selected, this being covered in Chapter 3.

From there on, Chapters 4 through 14 deal with all the other aspects of Practical Process Control, namely the controller(s), functions, actions and reactions, function combinations and various modes of operation. By way of introduction to the controller itself, the last sections of this chapter are introductions to the basic definitions of controller terms and types of control modes that are available.

1.3   Basic definitions and terms used in process control

Most basic process control systems consist of a control loop as shown in Figure 1.1, having four main components:

1. A measurement of the state or condition of a process
2. A controller calculating an action based on this measured value against a preset or desired value (setpoint)
3. An output signal resulting from the controller calculation, which is used to manipulate the process action through some form of actuator
4. The process itself reacting to this signal, and changing its state or condition.

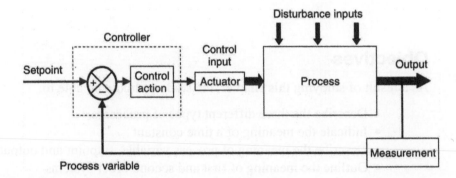

Figure 1.1
Block diagram showing the elements of a process control loop

As we will see in Chapters 2 and 3, two of the most important signals used in process control are called

1. Process variable or PV
2. Manipulated variable or MV.

In industrial process control, the PV is measured by an instrument in the field, and acts as an input to an automatic controller which takes action based on the value of it. Alternatively, the PV can be an input to a data display so that the operator can use the reading to adjust the process through manual control and supervision.

The variable to be manipulated, in order to have control over the PV, is called the MV. For instance, if we control a particular flow, we manipulate a valve to control the flow. Here, the valve position is called the MV and the measured flow becomes the PV.

In the case of a simple automatic controller, the Controller Output Signal (OP) drives the MV. In more complex automatic control systems, a controller output signal may drive the target values or reference values for other controllers.

The ideal value of the PV is often called the target value, and in the case of an automatic control, the term setpoint (SP) value is preferred.

1.4   Process modeling

To perform an effective job of controlling a process, we need to know how the control input we are proposing to use will affect the output of the process. If we change the input conditions we shall need to know the following:

- Will the output rise or fall?
- How much response will we get?

- How long will it take for the output to change?
- What will be the response curve or trajectory of the response?

The answers to these questions are best obtained by creating a mathematical model of the relationship between the chosen input and the output of the process in question. Process control designers use a very useful technique of block diagram modeling to assist in the representation of the process and its control system. The principles that we should be able to apply to most practical control loop situations are given below.

The process plant is represented by an input/output block as shown in Figure 1.2.

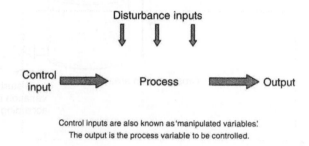

Control inputs are also known as 'manipulated variables.'
The output is the process variable to be controlled.

Figure 1.2
Basic block diagram for the process being controlled

In Figure 1.2 we see a controller signal that will operate on an input to the process, known as the MV. We try to drive the output of the process to a particular value or SP by changing the input. The output may also be affected by other conditions in the process or by external actions such as changes in supply pressures or in the quality of materials being used in the process. These are all regarded as disturbance inputs and our control action will need to overcome their influences as best as possible.

The challenge for the process control designer is to maintain the controlled process variable at the target value or change it to meet production needs, whilst compensating for the disturbances that may arise from other inputs. So, for example, if you want to keep the level of water in a tank at a constant height whilst others are drawing off from it, you will manipulate the input flow to keep the level steady.

The value of a process model is that it provides a means of showing the way the output will respond to the actions of the input. This is done by having a mathematical model based on the physical and chemical laws affecting the process. For example, in Figure 1.3 an open tank with cross-sectional area A is supplied with an inflow of water Q_1 that can be controlled or manipulated. The outflow from the tank passes through a valve with a resistance R to the output flow Q_2. The level of water or pressure head in the tank is denoted as H. We know that Q_2 will increase as H increases, and when Q_2 equals Q_1 the level will become steady.

The block diagram version of this process is drawn in Figure 1.4.

Note that the diagram simply shows the flow of variables into function blocks and summing points, so that we can identify the input and output variables of each block.

We want this model to tell us how H will change if we adjust the inflow Q_1 whilst we keep the outflow valve at a constant setting. The model equations can be written as follows:

$$\frac{dH}{dt} = \frac{Q_1 - Q_2}{A} \quad \text{and} \quad Q_2 = \frac{H}{R}$$

The first equation says the rate of change of level is proportional to the difference between inflow and outflow divided by the cross-sectional area of the tank. The second equation says the outflow will increase in proportion to the pressure head divided by the flow resistance R.

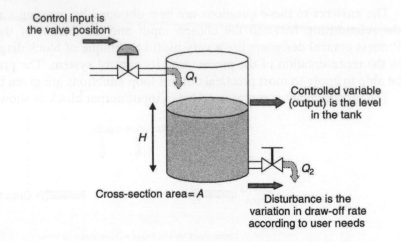

Figure 1.3
Example of a water tank with controlled inflow

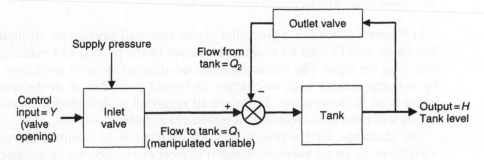

Figure 1.4
Elementary block diagram of tank process

Cautionary note: For turbulent flow conditions in the exit pipe and the valve, the effective resistance to flow R will actually change in proportion to the square root of the pressure drop so we should also note that $R = $ a constant $x \times H$. This creates a non-linear element in the model which makes things more complicated. However, in control modeling it is common practice to simplify the nonlinear elements when we are studying dynamic performance around a limited area of disturbance. So, for a narrow range of level we can treat R as a constant. It is important that this approximation is kept in mind because in many applications it often leads to problems when loop tuning is being set up on the plant at conditions away from the original working point.

The process input/output relationship is therefore defined by substituting for Q_2 in the linear differential equation

$$\frac{dH}{dt} = \frac{Q_1}{A} - \frac{H}{RA}$$

which is rearranged to a standard form as

$$(RA)\left(\frac{dH}{dt}\right) + H = RQ_1$$

When this differential equation is solved for H it gives

$$H = RQ_1\left(1 - e^{\frac{-t}{RA}}\right)$$

Using this equation we can show that if a step change in flow ΔQ_1 is applied to the system, the level will rise by the amount $\Delta Q_1 R$, by following an exponential rise vs time. This is the characteristic of a first order dynamic process and is very commonly seen in many physical processes. These are sometimes called capacitive and resistive processes, and include examples such as charging a capacitor through a resistance circuit (see Figure 1.5) and heating of a well-mixed hot water supply tank (see Figure 1.6).

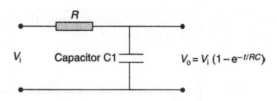

Figure 1.5
Resistance and capacitor circuit with first order response

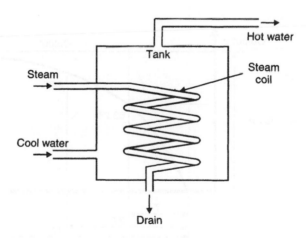

Figure 1.6
Resistance and capacitance effects in a water heater

1.5   Process dynamics and time constants

Resistance, capacitance and inertia are perhaps the most important effects in industrial processes involving heat transfer, mass transfer and fluid flow operations. The essential characteristics of first and second order systems are summarized below, and they may be used to identify the time constant and responses of many processes as well as mechanical and electrical systems. In particular, it should be noted that most process measuring instruments will exhibit a certain amount of dynamic lag, and this must be recognized in any control system application since it will be a factor in the response and in the control loop tuning.

1.5.1   First order process dynamic characteristics

The general version of the process model for a first order lag system is a linear first order differential equation:

$$T\frac{\mathrm{d}c(t)}{\mathrm{d}t}+c(t)=K_{\mathrm{p}}m(t)$$

Where
- T = the process response time constant
- K_{p} = the process steady-state gain (output change/input change)
- t = time
- $c(t)$ = process output response
- $m(t)$ = process input response.

The output of a first order process follows the step-change input with a classical exponential rise as shown in Figure 1.7.

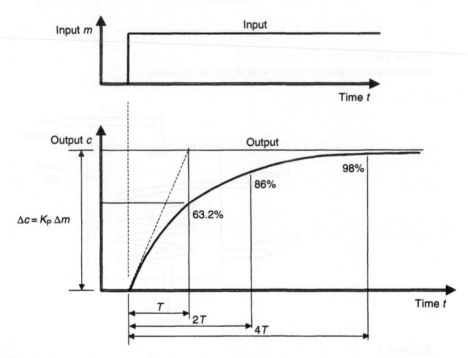

Figure 1.7
First order response

Important points to note: T is the time constant of the system and is the time taken to reach 63.2% of the final value after a step change has been applied to the system. After four time constants the output response has reached 98% of the final value that it will settle at.

$$K_{\mathrm{p}}\text{ is the steady-state gain}=\frac{\text{Final steady-state change in output}}{\text{Change in input}}$$

The initial rate of rise of the output will be K_{p}/T.

Application to the tank example

If we apply some typical tank dimensions to the response curve in Figure 1.7 we can predict the time that the tank-level example in Figure 1.3 will need to stabilize after a small step change around a target level H.

For example, suppose the tank has a cross-sectional area of 2 m^2 and operates at $H = 2$ m when the outflow rate is 5 m^3/h. The resistance constant R will be $H/Q_2 = 2 \text{ m}/5 \text{ m}^3/\text{h} = 0.4 \text{ h/m}^2$ and the time constant will be $AR = 0.8$ h. The gain for a change in Q_1 will also be R.

Hence, if we make a small corrective change at Q_1 of say 0.1 m^3/h the resulting change in level will be: $RQ_1 = 1 \times 0.4 = 0.4$ m, and the time to reach 98% of that change will be 3.2 h.

1.5.2   Resistance process

Now that we have seen how a first order process behaves, we can summarize the possible variations that may be found by considering the equivalent of resistance, capacitance and inertia type processes.

If a process has very little capacitance or energy storage the output response to a change in input will be instantaneous and proportional to the gain of the stage. For example, if a linear control valve is used to change the input flow in the tank example of Figure 1.3, the output flow will rise immediately to a higher value with a negligible lag.

1.5.3   Capacitance type processes

Most processes include some form of capacitance or storage capability, either for materials (gas, liquid, or solids) or for energy (thermal, chemical, etc.). Those parts of the process with the ability to store mass or energy are termed 'capacities'. They are characterized by storing energy in the form of potential energy; for example, electrical charge, fluid hydrostatic head, pressure energy and thermal energy.

The capacitance of a liquid- or gas-storage tank is expressed in area units. These processes are illustrated in Figure 1.8. The gas capacitance of a tank is constant and is analogous to electrical capacitance.

The liquid capacitance equals the cross-sectional area of the tank at the liquid surface; if this is constant then the capacitance is also constant at any head.

Using Figure 1.8 consider now what happens if we have a steady-state condition, where flow into the tank matches the flow out via an orifice or valve with flow resistance R. If we change the inflow slightly by ΔV the outflow will rise as the pressure rises until we have a new steady-state condition. For a small change we can take R to be a constant value. The pressure and outflow responses will follow the first order lag curve we have seen in Figure 1.7 and will be given by the equation $\Delta p = R \Delta V (1 - e^{-t/RC})$ and the time constant will be RC.

It should be clear that this dynamic response follows the same laws as those for the liquid tank example in Figure 1.3 and for the electrical circuit shown in Figure 1.5.

A purely capacitive process element can be illustrated by a tank with only an inflow connection such as shown in Figure 1.9. In such a process, the rate at which the level rises is inversely proportional to the capacitance and the tank will eventually flood. For an initially empty tank with constant inflow, the level c is the product of the inflow rate m and the time period of charging t divided by the capacitance of the tank C.

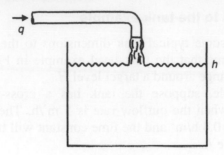

Liquid capacitance is defined by $C = \dfrac{dv}{dh}$ ft^2

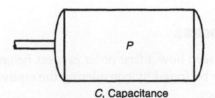

P

C, Capacitance

Gas capacitance is defined by

$$C = \frac{dv}{dp} = \frac{V}{nRT} \text{ ft}^2$$

Where
  v = weight of gas in vessel, lb.
  V = volume of vessel, ft^3
  R = Gas constant of a specific gas, ft/deg
  p = pressure, lb ft^2
  n = polytropic exponent is between
     1.0 and 1.2 for uninsulated tanks

Figure 1.8
Capacitance of a liquid or gas storage tank expressed in area units

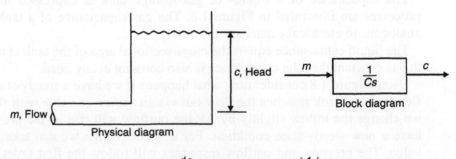

c, Head

m

$\dfrac{1}{Cs}$

c

Block diagram

m, Flow

Physical diagram

$$C\frac{dc}{dt} = m = (Cs)c = m \Rightarrow c = \left(\frac{1}{Cs}\right)m$$

Where
  C = capacitance
  c = output variable (head)
  t = time
  m = input variable (flow)
  $s = \dfrac{d}{dt}$ = differential operator

Figure 1.9
Liquid capacitance calculation; the capacitance element

1.5.4 Inertia type processes

Inertia effects are typically due to the motion of matter involving the storage or dissipation of kinetic energy. They are most commonly associated with mechanical systems involving moving components, but are also important in some flow systems in which fluids must be accelerated or decelerated. The most common example of a first order lag caused by kinetic energy build-up is when a rotating mass is required to change speed or when a motor vehicle is accelerated by an increase in engine power up to a higher speed, until the wind and rolling resistances match the increased power input.

For example consider a vehicle of mass M moving at $V = 60$ km/h, where the driving force F of the engine matches the wind drag and rolling resistance forces. If B is the coefficient of resistance, the steady state is $F = VB$, and for a small change of force ΔF the final speed change will be $\Delta V = \Delta F/B$.

The speed change response will be given by

$$\Delta V = \left(\frac{\Delta F}{B}\right) \times \left(1 - e^{\frac{-tB}{M}}\right)$$

This equation is directly comparable to the versions for the tank and the electrical *RC* circuit. In this case, the time constant is given by M/B. Obviously, the higher the mass of the vehicle the longer it will take to change speed for the same change in driving force. If the resistance to speed is high, the speed change will be small and the time constant will be shorter.

1.5.5 Second order response

Second order processes result in a more complicated response curve. This is due to the exchange of energy between inertia effects and interactions between first order resistance and capacitance elements. They are described by the following second order differential equation:

$$T^2 \frac{d^2 c(t)}{dt^2} + 2\xi T \frac{dc(t)}{dt} + c(t) = K_P m(t)$$

Where
T = the time constant of the second order process
ξ = the damping ratio of the system
K_p = the system gain
t = time
$c(t)$ = process output response
$m(t)$ = process input response.

The solutions to the equation for a step change in $m(t)$ with all initial conditions zero can be any one of a family of curves as shown in Figure 1.10. There are three broad classes of response in the solution, depending on the value of the damping ratio:

1. $\xi < 1.0$, the system is underdamped and overshoots the steady-state value. If $\xi < 0.707$, the system will oscillate about the final steady-state value.
2. $\xi > 1.0$, the system is overdamped and will not oscillate or overshoot the final steady-state value.
3. $\xi = 1.0$, the system is critically damped. In this state it yields the fastest response without overshoot or oscillation. The natural frequency of oscillation will be $\omega_n = 1/T$ and is defined in terms of the 'perfect' or 'frictionless'

situation where $\xi = 0.0$. As the damping factor increases, the oscillation frequency decreases or stretches out until the critical damping point is reached.

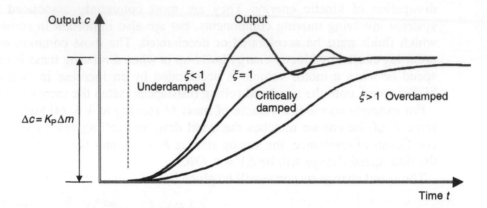

Figure 1.10
Step response of a second order system

For practical application in control systems the most common form of second order system is found wherever two first order lag stages are in series, in which the output of the first stage is the input to the second. As we shall see in Section 1.4.9 where the lags are modeled using transfer functions, the time constants of the two first order lags are combined to calculate the equivalent time constant and damping factor for their overall response as a second order system.

Important note: When a simple feedback control loop is applied to a first order system or to a second order system, the overall transfer function of the combined process and control system will usually be equivalent to a second order system. Hence, the response curves shown in Figure 1.10 will be seen in typical closed loop control system responses.

1.5.6   Multiple time constant processes

In multiple time constant processes, say where two tanks are connected in series, the process will have two or more two time lags operating in series. As the number of time constants increases, the response curves of the system become progressively more retarded and the overall response gradually changes into an S-shaped reaction curve as can be seen in Figure 1.11.

1.5.7   High order response

Any process that consists of a large number of process stages connected in series can be represented by a set of series-connected first order lags or transfer functions. When combined for the overall process, they represent a high order response, but very often one or two of the first order lags will be dominant or can be combined. Hence, many processes can be reduced to approximate first or second order lags, but they will also exhibit a dead time or transport lag as well.

1.5.8   Dead time or transport delay

For a pure dead-time process, whatever happens at the input is repeated at the output θd time units later, where θd is the dead time. This would be seen, for example, in a long pipeline if the liquid blend was changed at the input or the liquid temperature was changed at the input and the effects were not seen at the output until the travel time in the pipe has expired.

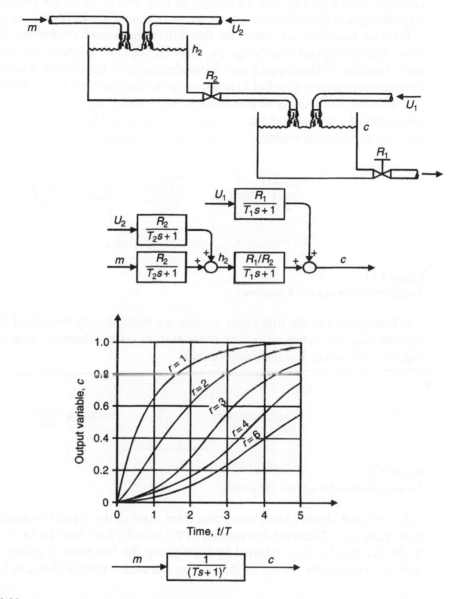

Figure 1.11
Response curves of processes with several time constants

In practice, the mathematical analysis of uncontrolled processes containing time delays is relatively simple, but a time delay, or a set of time delays, within a feedback loop tends to lend itself to very complex mathematics.

In general, the presence of time delays in control systems reduces the effectiveness of the controller. In well-designed systems the time delays (dead times) should be kept to the minimum.

1.5.9 Using transfer functions

In practice, differential equations are difficult to manipulate for the purposes of control system analysis. The problem is simplified by the use of transfer functions.

Transfer functions allow the modeling blocks to be treated as simple functions that operate on the input variable to produce the output variable. They operate only on changes from a steady-state condition, so they will show us the time response profile for step changes or disturbances around the steady-state working point of the process.

Transfer functions are based on the differential equations for the time response being converted by Laplace transforms into algebraic equations which can operate directly on the input variable. Without going into the mathematics of transforms, it is sufficient to note that the transient operator (symbol S) replaces the differential operator such that d(variable)/d$t = S$.

A transfer function is abbreviated as $G(s)$ and it represents the ratio of the Laplace transform of a process output $C(s)$ to that of an input $M(s)$, as shown in Figure 1.12. From this, the simple relationship $C(s) = G(s)M(s)$ is obtained.

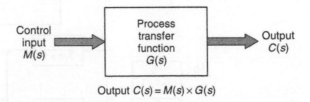

Output $C(s) = M(s) \times G(s)$

Figure 1.12
Transfer function in a block diagram

When applied to the first order system, we have already described the transfer function representing the action of a first order system on a changing input signal, as shown in Figure 1.13, where T is the time constant.

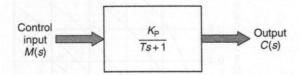

Figure 1.13
Transfer function for a first order process

As we have already seen, many processes involve the series combination of two or more first order lags. These are represented in the transfer function blocks as seen in Figure 1.14. If the two blocks are combined by multiplying the functions together, they can be seen to form a second order system as shown here and as described in Section 1.4.5.

Two first order lags in series

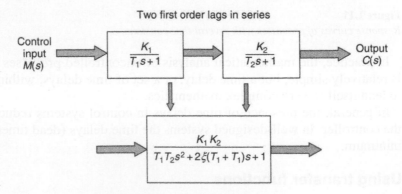

Figure 1.14
Two lags in series combine to produce a second order system

Block diagram modeling of the control system proceeds in the same manner as for the process, and is shown by adding the feedback controller as one or more transfer function blocks. The most useful rule for constructing the transfer function of a feedback control loop is shown in Figure 1.15.

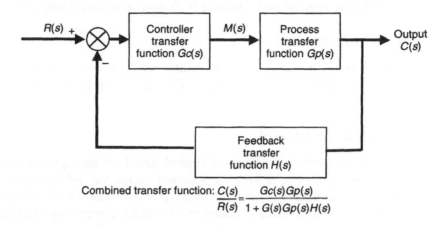

Combined transfer function: $\dfrac{C(s)}{R(s)} = \dfrac{Gc(s)Gp(s)}{1 + G(s)Gp(s)H(s)}$

Figure 1.15
Block diagram and transfer function for a typical feedback control system

The feedback transfer function $H(s)$ (typically the sensor response) and the controller transfer function $Gc(s)$ are combined in the model to give an overall transfer function that can be used to calculate the overall behavior of the controlled process.

This allows the complete control system working with its process to be represented in an equation known as the closed loop transfer function. The denominator of the right-hand side of this equation is known as the open loop transfer function. You can see that if this denominator becomes equal to zero, the output of the process approaches infinity and the whole process is seen to be unstable. Hence, control engineering studies place great emphasis on detecting and avoiding the condition where the open loop transfer function becomes negative and the control system becomes unstable.

1.6   Types or modes of operation of process control systems

There are five basic forms of control available in process control. These are:

1. On–off
2. Modulating
3. Open loop
4. Feedforward
5. Closed loop.

The next five sections (1.6.1–1.6.5) examine each of these in turn.

1.6.1   On–off control

The most basic control concept is on–off control, as found in a modern iron in our households. This is a very crude form of control, which nevertheless should be considered as a cheap and effective means of control if a fairly large fluctuation of the PV is acceptable.

The wear and tear of the controlling element (solenoid valve etc.) needs special consideration. As the bandwidth of fluctuation of a PV increases, the frequency of switching (and thus wear and tear) of the controlling element decreases.

1.6.2   Modulating control

If the output of a controller can move through a range of values, we have modulating control. It is understood that modulating control takes place within a defined operating range (with an upper and lower limit) only.

Modulating control can be used in both open and closed loop control systems.

1.6.3   Open loop control

We have open loop control if the control action (Controller Output Signal OP) is not a function of the PV or load changes. The open loop control does not self-correct when these PVs drift.

1.6.4   Feedforward control

Feedforward control is a form of control based on anticipating the correct manipulated variables necessary to deliver the required output variable. It is seen as a form of open loop control as the PV is not used directly in the control action. In some applications, the feedforward control signal is added to a feedback control signal to drive the MV closer to its final value. In other more advanced control applications, a computer-based model of the process is used to compute the required MV and this is applied directly to the process as shown in Figure 1.16.

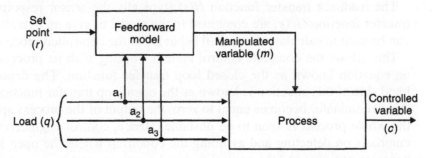

Figure 1.16
A model based feedforward control system

For example, a typical application of this type of control is to incorporate this with feedback – or closed loop control. Then the imperfect feedforward control can correct up to 90% of the upsets, leaving the feedback system to correct the 10% deviation left by the feedforward component.

1.6.5   Closed loop or feedback control

We have a closed loop control system if the PV, the objective of control, is used to determine the control action. The principle is shown in Figure 1.17.

The idea of closed loop control is to measure the PV; compare this with the SP which is the desired or target value; and determine a control action which results in a change of the OP value of an automatic controller.

In most cases, the ERROR (ERR) term is used to calculate the OP value.

$$ERR = PV - SP$$

If ERR = SP – PV has to be used, the controller has to be set for REVERSE control action.

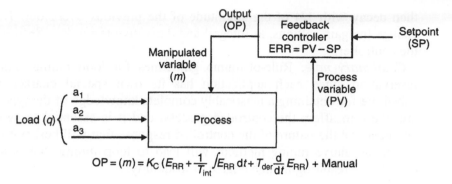

Figure 1.17
The feedback control loop

1.7   Closed loop controller and process gain calculations

In designing and setting up practical process control loops, one of the most important tasks is to establish the true factors making up the loop gain and then to calculate the gain.

Typically, the constituent parts of the entire loop will consist of a minimum of four functional items:

1. Process gain: $(K_P) = \Delta PV/\Delta MV$
2. Controller gain: $(K_C) = \Delta MV/\Delta E$
3. The measuring transducer or sensor gain (refer to Chapter 2), K_S and
4. The valve gain K_V.

The total loop gain is the product of these four operational blocks.

For simple loop tuning, two basic methods have been in use for many years. The Zeigler and Nichols method is called the 'ultimate cycle method' and requires that the controller should first be set up with proportional-only control. The loop gain is adjusted to find the ultimate gain, K_u. This is the gain at which the MV begins to sustain a permanent cycle. For a proportional-only controller the gain is then reduced to 0.5 K_u. Therefore for this tuning the loop gain must be considered in terms of the sum of the four gains given above, and the tuning condition is given by the following equation:

$$K_{LOOP} = (K_C \times K_P) = \left(\frac{\Delta MV}{\Delta E} \times \frac{\Delta PV}{\Delta MV} = \frac{\Delta PV}{\Delta E} \right) \times K_S \times K_V = 0.5 \times K_u$$

Normally, only the controller gain can be changed, but it remains very important that the other gain components be recognized and calculated. In particular, the valve gain and process gain may change substantially with the working point of the process, and this is the cause of many of the tuning problems encountered on process plants.

Other gain settings are used in the Zeigler and Nichols method for PI and PID controllers to ensure stability when integral and derivative actions are added to the controller. See the next section (Section 1.8) for the meaning of these terms.

The alternative tuning method is known as the 1/4 damping method. This suggests that the controller gain should be adjusted to obtain an under-damped overshoot response having a quarter amplitude of the initial step change in setpoint. Subsequent oscillations then decay with 1/4 of the amplitude of the previous overshoot. This method does not change the gain settings, as integral and derivative terms (see Section 1.8) are added into the controller.

Cautionary note: Rule-of-thumb guidelines for loop tuning should be treated with reservation since each application has its own special characteristics. There is no substitute for obtaining a reasonably complete knowledge of the type of disturbances that are likely to affect the controlled process, and it is essential to agree with the process engineers on the nature of the controlled response that will best suit the process. In some cases, the above tuning methods will lead to loop tuning that is too sensitive for the conditions, resulting in high degree of instability.

1.8   Proportional, integral and derivative control modes

Most closed loop controllers are capable of controlling with three control modes, which can be used separately or together:

1. Proportional control (P)
2. Integral or reset control (I)
3. Derivative or rate control (D).

The purpose of each of these control modes is as follows:

Proportional control   This is the main and principal method of control. It calculates a control action proportional to the ERROR. Proportional control cannot eliminate the ERROR completely.

Integral control (reset)   This is the means to eliminate the remaining ERROR or OFFSET value, left from the proportional action, completely. This may result in reduced stability in the control action.

Derivative control (rate)   This is sometimes added to introduce dynamic stability to the control LOOP.

Note: The terms 'reset' for integral and 'rate' for derivative control actions are seldom used nowadays.

Derivative control has no functionality of its own.

The only combinations of the P, I and D modes are as follows:

- P   For use as a basic controller
- PI   Where the offset caused by the P mode is removed
- PID   To remove instability problems that can occur in PI mode
- PD   Used in cascade control; a special application
- I   Used in the primary controller of cascaded systems.

1.9   An introduction to cascade control

Controllers are said to be 'in cascade' when the output of the first or primary controller is used to manipulate the SD of another or secondary controller. When two or more controllers are cascaded, each will have its own measurement input or PV, but only the primary controller can have an independent SP and only the secondary, or the most down-stream, controller has an output to the process.

Cascade control is of great value where high performance is needed in the face of random disturbances, or where the secondary part of a process contains a significant time lag or has nonlinearity.

The principal advantages of cascade control are the following:

- Disturbances occurring in the secondary loop are corrected by the secondary controller before they can affect the primary, or main, variable.
- The secondary controller can significantly reduce phase lag in the secondary loop, thereby improving the speed or response of the primary loop.
- Gain variations due to nonlinearity in the process or actuator in the secondary loop are corrected within that loop.
- The secondary loop enables exact manipulation of the flow of mass or energy by the primary controller.

Figure 1.18 shows an example of cascade control where the primary controller TC is used to measure the output temperature T_2, and compare this with the SP value of the TC; and the secondary controller, FC, is used to keep the fuel flow constant against variables like pressure changes.

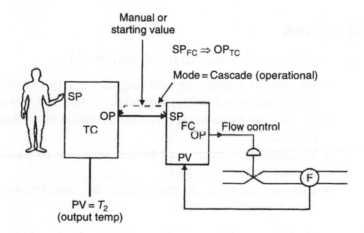

Figure 1.18
An example of cascade control

The primary controller's output is used to manipulate the SP of the secondary controller, thereby changing the fuel feed rate to compensate for temperature variations of T_2 only. Variations and inconsistencies in the fuel flow rate are corrected solely by the secondary controller – the FC controller.

The secondary controller is tuned with a high gain to provide a proportional (linear) response to the range, thereby removing any nonlinear gain elements from the action of the primary controller.

2
Process measurement and transducers

2.1   Objectives

At the conclusion of this chapter, the student should:

- Be able to explain the meaning of the terms accuracy, precision, sensitivity, resolution, repeatability, rangeability, span and hysteresis
- Be able to make an appropriate selection of sensing devices for a particular process
- Describe the sensors used for measurement of temperature, pressure, flow and liquid level
- List the methods of minimizing the interference effects of noise on our instrumentation system.

2.2   The definition of transducers and sensors

A transducer is a device that obtains information in the form of one or more physical quantities and converts this into an electrical output signal. Transducers consist of two principle parts, a primary measuring element referred to as a sensor, and a transmitter unit responsible for producing an electrical output that has some known relationship to the physical measurement as the basic components.

In more sophisticated units, a third element may be introduced which is quite often microprocessor based. This is introduced between the sensor and the transmitter part of the unit and has amongst other things, the function of linearizing and ranging the transducer to the required operational parameters.

2.3   Listing of common measured variables

In descending order of frequency of occurrence, the principal controlled variables in process control systems comprise:

- Temperature
- Pressure
- Flow rate
- Composition
- Liquid level.

Sections 2.4 through to 2.6 of this chapter list and describe these different types of transducers, ending with a methodology of selecting sensing devices.

2.4 The common characteristics of transducers

All transducers, irrespective of their measurement requirements, exhibit the same characteristics such as range, span, etc. This section explains and demonstrates the interpretation of the most common of these characteristics.

2.4.1 Dynamic and static accuracy

The very first, and most common term *accuracy* is also the most misused and least understood.

It is nearly always quoted as 'this instrument is ±X% accurate', when in fact it should be stated as 'this instrument is ±X% inaccurate'. In general, accuracy can be best described as how close the measurement's indication is to the absolute or real value of the process variable. In order to obtain a clear understanding of this term, and all of the other ones that are associated with it, the term *error* should first be defined.

The definition of error in process control

Error means a mistake or transgression, and is the difference between a perfect measurement and what was actually measured at any point, time and direction of process movement in the process measuring range.

There are two types of accuracy, *static* or *steady-state* accuracy and *dynamic* accuracy.

1. *Static accuracy* is the closeness of approach to the true value of the variable when that true value is constant.
2. *Dynamic accuracy* is the closeness of approach of the measurement when the true value is changing, remembering that a measurement lag occurs here, that is to say, by the time the measurement reading has been acted upon, the actual physical measured quantum may well have changed.

In addition to the term *accuracy*, a sub-set of terms appear, these being *precision*, *sensitivity*, *resolution*, *repeatability* and *rangeability* all of which have a relationship and association with the term *error*.

2.4.2 Precision

Precision is the accuracy with which repeated measurements of the same variable can be made under identical conditions.

In process control, precision is more important than accuracy, i.e. it is usually preferable to measure a variable precisely than it is to have a high degree of absolute accuracy. The difference between these two properties of measurement is illustrated in Figure 2.1.

Using a fluid as an example, the dashed curve represents the actual or real temperature. The upper measurement illustrates a precise but inaccurate instrument while the lower measurement illustrates an imprecise but more accurate instrument. The first instrument has the greater error, the latter has the greater drift.

(*Drift*: An undesirable change in the output to input relationship over a period of time.)

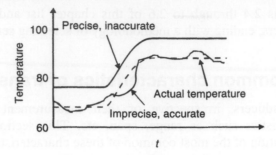

Figure 2.1
Accuracy vs precision related to a typical temperature measurement

2.4.3 Sensitivity

Generally, sensitivity is defined as the amount of change in the output signal from a transducer's transmitting element to a specified change in the input variable being measured, i.e. it is the ratio of the output signal change to the change in the measured variable and is a steady-state ratio or the steady-state gain of the element. So, the greater the output signal change from the transducer's transmitter for a given input change, the greater the sensitivity of the measuring element.

Highly sensitive devices, such as thermistors, may change resistance by as much as 5% per °C, while devices with low sensitivity, such as thermocouples, may produce an output voltage which changes by only 5 µV (5×10^{-6} V) per °C.

The second kind of sensitivity important to measuring systems is defined as the smallest change in the measured variable which will produce a change in the output signal from the sensing element.

In many physical systems, particularly those containing levers, linkages and mechanical parts, there is a tendency for these moving parts to stick and to have some free play.

The result of this is that small input signals may not produce any detectable output signal. To attain high sensitivity, instruments need to be well-designed and well-constructed. The control system will then have the ability to respond to small changes in the controlled variable; it is sometimes known as *resolution*.

2.4.4 Resolution

Precision is related to *resolution*, which is defined as the smallest change of input that results in a significant change in transducer output.

2.4.5 Repeatability

The closeness of agreement between a number of consecutive measurements of the output for the same value of input under identical operating conditions, approaching from the same direction for full range transverses is usually expressed as repeatability in percent of span. It does not include hysteresis.

2.4.6 Rangeability

This is the region between stated upper and lower range values of which the quantity is measured. Unless otherwise stated, input range is implied.

Example:

If the range is stated as 50–320 °C then the range is quoted as 50–320 °C.

2.4.7 Span

Span should not be confused with rangeability, although the same points of reference are used. Span is the *Algebraic* difference between the *upper and lower range values*.

Example:

If the range is stated, as in Section 2.4.6, as 50–320 °C then the span is 320–50 = 270 °C.

2.4.8 Hysteresis

This is a dynamic measurement, and shows as the dependency of an output value, given for an excursion of the input, as compared with the history of prior excursions and the direction of the transverse.

Example:

If an input into a system is moved between 0 and 100% and the resultant output recorded and then the input is returned back to 0%, again with the output recorded the difference between the two values, $0\% \Rightarrow 100\% \Rightarrow 0\%$, as recorded, gives the hysteresis value of the system at all points in its range. Repetitive tests must be done under identical conditions.

2.5 Sensor dynamics

Process dynamics have been discussed in Chapter 1, and these same factors will apply to a sensor making it important to gain an understanding of sensor dynamics. The speed of response of the primary measuring element is often one of the most important factors in the operation of a feedback controller. As process control is continuous and dynamic, the rate at which the controller is able to detect changes in the process will be critical to the overall operation of the system.

Fast sensors allow the controller to function in a timely manner, while sensors with large time constants are slow and degrade the overall operation of the feedback loop. Due to their influence on loop response, the dynamic characteristics of sensors should be considered in their selection and installation.

2.6 Selection of sensing devices

A number of factors must be considered before a specific means of measuring the process variable (PV) can be selected for a particular loop:

- The normal range over which the PV may vary, and if there are any extremes to this
- The accuracy, precision and sensitivity required for the measurement
- The sensor dynamics required
- The reliability that is required
- The costs involved, including installation and operating costs as well as purchase costs
- The installation requirements and problems, such as size and shape restraints, remote transmission, corrosive fluids, explosive mixtures, etc.

2.7   Temperature sensors

Temperature is the most common PV measured in process control. Due to the vast temperature range that needs to be measured (from absolute zero to thousands of degrees) with spans of just a few degrees and sensitivities down to fractions of a degree, there is a vast range of devices that can be used for temperature measurements.

The five most common sensors; *thermocouples*, *resistance temperature detectors or RTDs*, *thermistors*, *IC sensors* and *radiation pyrometers* have been selected for this chapter as they illustrate most of the application, range, accuracy and linearity aspects that are associated with temperature measurements.

2.7.1   Thermocouples

Thermocouples cover a range of temperatures, from –262 to +2760 °C and are manufactured in many materials, are relatively cheap, have many physical forms, all of which make them a highly versatile device.

Thermocouples suffer from two major problems that cause errors when applying them to the process control environment.

1. The first is the small voltages generated by them, for example a 1 °C temperature change on a platinum thermocouple results in an output change of only 5.8 µV = $(5.8 \times 10^{-6}$ V).
2. The second is their non-linearity, requiring polynomial conversion, look up tables or related calibration to be applied to the signaling and controlling unit (see Figure 2.2).

Ranges of six types of common thermocouples

Metal Composition	Temperature Span	Seebeck Coefficient
K Chromel vv alumel	–190 to +1371 °C	40 µV/°C
J Iron vv constantan	–190 to +760 °C	50 µV/°C
T Copper vv constantan	–190 to +760 °C	50 µV/°C
E Chromel vv constantan	–190 to +1472 °C	60 µV/°C
S Platinum vv 10% rhodium / platinum	0 to +1760 °C	10 µV/°C
R Platinum vv 13% rhodium / platinum	0 to +1670 °C	11 µV/°C

Table 2.1
Thermocouple types, temperature range and value of the seebeck effect

Principles of thermocouple operation

A thermocouple could be considered as a heat-operated battery, consisting of two different types of homogeneous (of the same kind and nature) metal or alloy wires joined together at one end of the measuring point and connected usually via special compensating cable, to some form of measuring instrument. At the point of connection to the measuring device a second junction is formed, called the reference or cold junction, which completes the circuit.

The Peltier and Thomson effects on thermocouple operation

The *Peltier effect* is the cause of the emfs generated at every junction of dissimilar metals in the circuit. This effect involves the generation or absorption of heat at the junction as current flows through it and temperature is dependent on current flow direction.

The *Thomson effect*, where a second emf can also be generated along the temperature gradient of a single homogeneous wire can also contribute to measurement errors. It is essential that all the wire in a thermocouple measuring circuit is homogeneous as then the emfs generated will be dependant solely on the types of material used. Any thermal emfs generated in the wire when it passes through temperature gradients will also be canceled from one to the other.

Additionally, if both junctions of a homogeneous metal are held at the same temperature, the metal will not contribute additional emfs to the circuit. It follows then that if all junctions in the circuit are held at a constant temperature, except the measuring one, measurement can be made of the hot, or measuring, junction value against the constant value or cold junction reference value.

Reference or cold junction compensation

As described in Section 2.7.1.3, we have to ensure that all the junctions in the measuring circuit, with the exception of the one being used for the actual process measurement, must either:

- Be held at a constant known temperature, usually 0 °C, and called a 'Cold Junction'
- Or the temperature of these junctions should be measured and the measuring instrument takes this into consideration when calculating its final output.

Both methods are commonly used (Figure 2.2); the first one, the cold junction, utilizes an isothermal block held at a known temperature and in which the connections from the thermocouple wires to copper wires are made. The second method is to measure the temperature, usually by a thermistor, at the point of copper to thermocouple connections, feeding this value into the measuring system and have that calculate a corrected output.

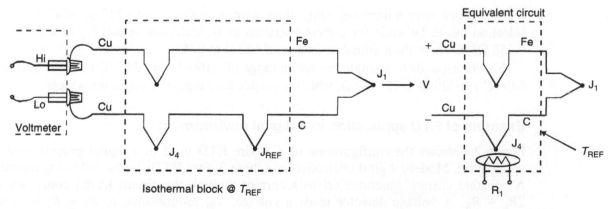

Figure 2.2
Thermocouple cold junction and reference junction circuit examples

2.7.2 Resistance temperature detectors or RTDs

In the same year as the discovery of the thermocouple by Thomas Seebeck, Sir Humphry Davy noted the temperature/resistivity dependence of metals, but it was H C Meyers who developed the first RTD in 1932.

Construction of RTDs

RTDs consist of a platinum or nickel wire element encased in a protective housing having, in the case of the platinum version a base resistance of 100 Ω at 0 °C and the nickel type a resistance of 1000 Ω, again at 0 °C (Figure 2.3).

They come packaged in either 2, 3 or 4 wire versions, the 3 and 4 wire being the most common. Two wire versions can be very inaccurate as the lead resistance is in series with the measuring circuit, and the measuring element relies on resistance change to indicate the temperature change.

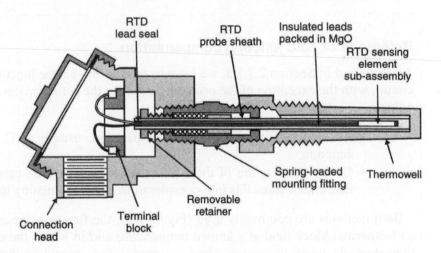

Figure 2.3
Construction of RTD

Range sensitivity and spans of RTDs

RTDs operate over a narrower range than thermocouples, from –247 to +649 °C. Span selection has to be made for correct operation as typically the sensitivity of a PT100 is 0.358 Ω/ °C about the nominal resistance of 100 Ω at 0 °C.

This corresponds to a single resistance range of (100–88 = –247 °C to 100 + 232 Ω = 649 °C) resulting in 12–332 Ω, which is outside the range of a single transducer.

Example of RTD application in a digital environment

Figure 2.4 shows the configuration of a 3-wire RTD used in a digital process control application. Modern digital controllers use these 3-wire RTDs in the following manner: A constant current generator drives a current through the circuit [A–C] consisting of $2R_L + R_X$. A voltage detector reads a voltage, V_B, proportional to $R_X + R_L$ between points [B and C] and a second voltage V_A which is proportional to $R_X + 2R_L$ between points [A and C].

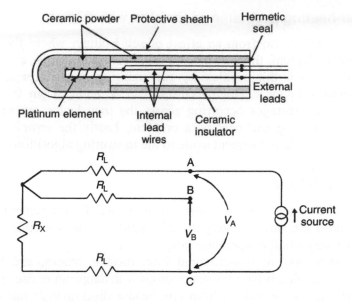

Figure 2.4
3-Wire RTD configuration for a digital system

As $V_A - V_B$ is proportional to R_L so $V_B - V_A - V_B$ is proportional to R_X where:

- R_L = The resistance of each of the three RTDs leads
- R_X = The measuring element of the RTD
- V_A = The voltage supply to the RTDs measuring element R_X from the constant current source
- V_B = The final measured voltage, or output from the RTD (3-wire version).

The measurements are made sequentially, digitized and stored until differences can be computed. RTDs are reasonably linear in operation, see Figure 2.5, but this depends to a great extent on the area of operation being used within the total span of the particular transducer in question.

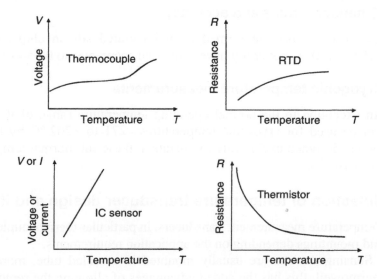

Figure 2.5
Characteristics of thermocouples, RTDs IC and thermistor temperature sensors

Self-heating problems associated with RTDs

RTDs suffer also from an effect of self-heating, where the excitation current heats the sensing element, thereby causing an error, or temperature offset. Modern digital systems can overcome this problem by energizing the transducer just before a reading is taken. Alternatively the excitation current can be reduced but this is at the expense of lower measuring voltages occurring across the transducers output, and subsequently induced electrical noise can become a problem. Lastly the error caused by self-heating can be calculated and adjustment made to the measuring algorithms.

2.7.3   Thermistors

These elements are the most sensitive and fastest temperature measuring devices in common use; unfortunately the price paid for this is terrible non-linearity (see Figure 2.5) and a very small temperature range.

Thermistors are manufactured from metallic oxides, and have a negative temperature coefficient, that is their resistance drops with temperature rise. They are also manufactured in almost any shape and size from a pin head to disks up to 25 mm diameter × 5 mm thickness.

Thermistor values, range and sensitivity

Most thermistors have a nominal quoted resistance of about 5000 Ω and because of their sensitivity, this base resistance is quoted at a specific temperature, reference having to be made to the relative type in the manufacturer's published specifications.

Thermistor values can change by as much as 200 $\Omega/°C$ which, in this case would give a maximum range of only +25 °C from the quoted base temperature.

2.7.4   IC sensors

Integrated circuitry sensors have only recently began to make their presence felt in the process control world. As such they are still limited in the variability of shape, size and packaging that is advisable. Their main advantages are their low cost (below $10.00) along with their linear and high output signals.

IC sensor ranges and accuracy

As these sensors are formed from integrated silicon chips, their range is limited to –55 to +150 °C but easily have calibrated accuracies to 0.05–0.1°C.

Cryogenic temperature measurements

An exception to the normal operating temperature range of IC sensors is a version that can be used for cryogenic temperatures –271 to +202 °C by the application of special diodes designed exclusively to operate at these sub-normal temperatures (absolute zero = –273.16 °C).

2.7.5   Selection of temperature transducer design and thermowells

Temperature measurement transducers, in particular thermocouples, need different housings and mountings depending on the application requirements.

Sensing devices are usually mounted in a sealed tube, more commonly known as a thermowell; this has the added advantages of allowing the removal or replacement of the sensing device without opening up the process tank or piping. Thermowells need to be considered when installing temperature-sensing equipment. The length of the thermowell needs to be sized for the temperature probe.

Consideration of the thermal response needs to be taken into account. If a fast response is required, and the sensor probe already has adequate protection, then a thermowell may impede system performance and response time. Note that when a thermowell is used, the response time is typically doubled.

Thermowells can provide added protection to the sensing equipment, and can also assist in maintenance and period calibration of equipment.

Thermopaste assists in the fast and effective transfer of thermal dynamics from the process to the sensing element. Application and maintenance of this material needs to be considered. Regular maintenance and condensation can affect the operation of the paste.

Figure 2.6 shows the three typical designs of thermocouple probes:

1. Open ended; subject to damage and should not be used in a hostile environment
2. Sealed and both thermally and electrically isolated from the outside world
3. Sealed but with thermal (and/or electrical) connection to the outside world.

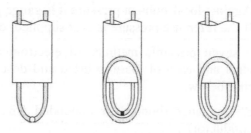

Figure 2.6
Sectional views of three typical thermocouple probes

2.7.6   Radiation pyrometers

At the other end of the scale is the requirement to measure high temperatures up to 4000 °C or more. Total radiation pyrometers operate by measuring the total amount of energy radiated by a hot body. Their temperature range is 0–3890 °C.

The infrared (IR) pyrometer is rapidly replacing this older type of measurement, and these work by measuring the dominant wavelength radiated by a hot body. The basis of this is in the fact that as temperature increases the dominant wavelength of hot body radiation gets shorter.

Developments in infrared optical pyrometry

Two recent developments in the world of pyrometry that should be mentioned are the utilization of lasers and fiber optics.

Lasers are used to automatically correct errors occurring due to changes in surface emissivity as the object's temperature changes.

Fiber optics can focus the temperature measurements on inaccessible or unfriendly areas. Some of these units are capable of very high accuracy, typically 0.1% at 1000 °C and can operate from 500 up to 2000 °C. Multi-plexing of the optics is also possible, reducing costs in multi-measuring environments.

2.8   Pressure transmitters

Pressure is probably the second most commonly used and important measurement in process control. The most familiar pressure measuring devices are manometers and dial gauges, but these require a manual operator.

For use in process control, a pressure measuring device needs a pressure transmitter that will produce an output signal for transmission, e.g. an electric current proportional to the pressure being measured. A transmitter typically that produces an output of a 4–20 mA signal is rugged and can be used in flammable or hazardous service.

2.8.1   Terms of pressure reference

Pressure is defined as force per unit area and may be expressed in units of newtons per square meter, millimeters of mercury, atmospheres, bars or torrs.

There are three common references against which it can be measured:

1. If measured against a vacuum, the measured pressure is called absolute pressure
2. Against local ambient pressure it is gauge pressure
3. If the reference pressure is user supplied, differential pressure is measured.

There are seven principle methods of electronically measuring pressure for use in process control and each of these is listed and described under its numeric heading, in principle detail below:

1. Strain gauge (bonded or unbonded wire or foil, bonded or diffused semi-conductor)
2. Capacitive
3. Potentiometric
4. Resonant wire
5. Piezoelectric
6. Magnetic (inductive and reluctive)
7. Optical.

2.8.2   Strain gauge

In process control applications, one of the most common ways to measure pressure is using a strain gauge sensor. There are two basic types of strain gauge, bonded and unbonded, each utilizing wire or foil, but both working in the same electrical manner. A thin wire (or foil strip), usually made from chrome–nickel alloys and sometimes platinum, is subjected to stretching, and hence its resistance increases as its length increases.

Strain gauges are commonly made using a thin metal wire or foil that is only a few micrometers thick, so they can also be known as thin film-based bonded pressure sensors.

The unbonded strain gauge consists of open wire wound round two parallel mounted posts which are flexed or pulled apart, hence imparting a stretching dynamic to the wire, reducing its resistance, these being physically much larger units (see Figure 2.7).

Composition of strain gauges

Bonded strain gauges are the most common type in use. They comprise an insulated bonded sandwich usually made of two sheets of paper, with the gauge wire laid in a specific pattern between them. Strain gauge wires of less than 0.001 in. (0.025 mm) diameter are used as they have a surface area several thousand times greater than the cross-sectional area.

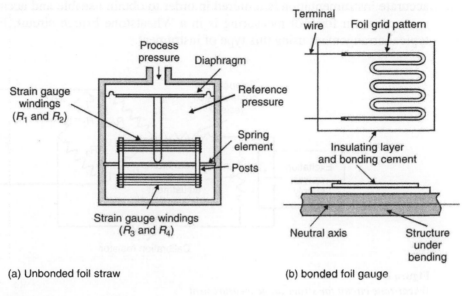

(a) Unbonded foil straw

(b) bonded foil gauge

Figure 2.7
Composition of bonded and unbonded strain gauges

Foil gauges have been commercially made where the foil thickness can be as low as 0.0001 in. (0.0025 mm).

Semiconductor types are available which have sensitivities close to one hundred times greater than the wire types.

Strain gauge sensitivity and gauge factor

The ratio of the percentage change in resistance to the percentage change in length is a measure of the sensitivity (S) of the gauge:

$$S = \frac{\left(\dfrac{\Delta R}{R}\right)}{\left(\dfrac{\Delta L}{L}\right)}$$

Where
  L is the initial length of the wire or foil
  R is the specific resistivity in the unstrained position.

Many things affect the axial strain and gauge resistance, such as the geometry of the wire or foil in the gauge, direction of strain. This is expressed by the constant called gauge factor or *GF*:

$$GF = \frac{\left(\dfrac{\Delta R}{R}\right)}{\text{Axial strain}}$$

Typically, four strain gauges are bonded to a metal or plastic flexible diaphragm and connected into a Wheatstone bridge circuit to yield an electrical signal proportional to the strain caused by the displacement of the diaphragm by the pressure applied to it.

The changes of resistance in a strain gauge are very small and as such precise and accurate instrumentation is required in order to obtain useable and accurate readings. The most common form of measuring is in a Wheatstone bridge circuit. Figure 2.8 shows a typical arrangement using this type of instrument.

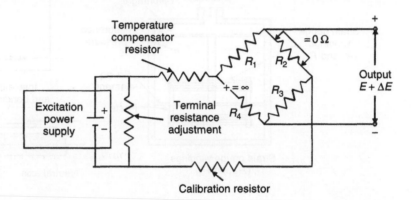

Figure 2.8
Wheatstone circuit for strain gauge measurement

Effects of temperature (ref. Section 2.7.2.3) on the gauge's resistance is minimal due to their influence on the output being subtractive.

The active gauges are located on opposite arms of the bridge, making effects additive; the other two gauges are for compensation and are either 'dummy' gauges or resistors with equal resistance to the active gauges.

This circuit is suited to both static and dynamic strain measurements, the output is, however, a differential output and care must be exercised in 'grounding' any part of this circuit.

An alternative arrangement, called a ballast or potentiometric circuit is arrived at by removing R_2 and making the value of R_3 equal to 0 in Figure 2.8.

Unfortunately this circuit is best used for dynamic sensing only; however one of the signal leads can now be grounded or electrically referenced to 0 V.

Measurement errors for strain gauges

A number of errors exist when strain is measured using the Wheatstone bridge arrangement. Typical of these are:

- Gauge factor uncertainty (typically 1%)
- Bridge non-linearity (typically 1%). This is a result of the assumption that the change in strain gauge resistance is very small compared to the nominal gauge resistance
- Matching of compensation resistors to the strain gauge (typically 0.5%)
- Measurement errors caused by the accuracy and resolution of the measuring device and lead resistances
- Temperature effects: Resistance varies with changes from the temperature at which a bridge is calibrated
- Self-heating of gauges.

Strain gauge pressure transducer specifications

Strain gauge elements can detect absolute gauge, and differential spans from 30 in. H$_2$O to upwards of 200 000 psig (7.5 kPa–1400 MPa). Their inaccuracy is around 0.2% and 0.5% of span. Units are available to work in the temperature range of –50 upto +120 °C with special units going to 320 °C.

2.8.3 Vibrating wire or resonant wire transducers

This type of sensor consists of an electronic oscillator circuit, which causes a wire to vibrate at its natural frequency when under tension. The principle is similar to that of a guitar string (Figure 2.9). A thin wire inside the sensor is kept in tension, with one end fixed and the other attached to a diaphragm. As pressure moves the diaphragm, the tension on the wire changes thereby changing its resonant vibration frequency.

These frequency changes are a direct consequence of pressure changes and as such is detected and shown as pressure.

The frequency can be sensed as digital pulses from an electromagnetic pickup or sensing coil. An electronic transmitter would then convert this into an electrical signal suitable for transmission.

This type of pressure measurement can be used for differential, absolute or gauge installations. Absolute pressure measurement is achieved by evacuating the low pressure diaphragm. A typical vacuum pressure for such a case would be about 0.5 Pa.

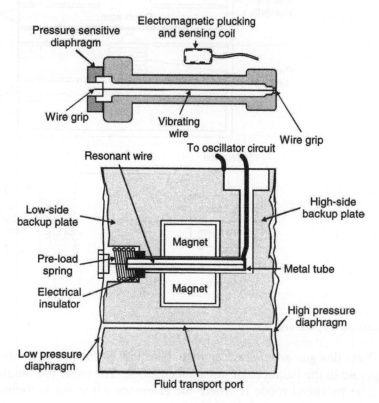

Figure 2.9
Resonant wire transducer

Transducer advantages

- Good accuracy and repeatability
- Stable
- Low hysteresis
- High resolution
- Absolute, gauge or differential measurement.

Transducer limitations, or disabilities

- Sensitivity to vibrations
- Temperature variations require temperature compensation within the sensor, this problem limits the sensitivity of the device
- The output generated is non-linear which can cause continuous control problems
- This technology is seldom used any more. Being an older technology, it is typically found with analog control circuitry.

2.8.4   Capacitance

Capacitive pressure measurement involves sensing the change in capacitance that results from the movement of a diaphragm (Figure 2.10). The sensor is energized electrically with a high frequency oscillator. As the diaphragm is deflected due to pressure changes, the relative capacitance is measured by a bridge circuit.

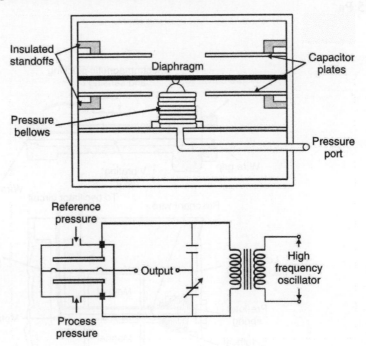

Figure 2.10
Capacitance pressure detector

Two designs are quite common. The first is the two-plate design and is configured to operate in the balanced or unbalanced mode. The other is a single capacitor design.

The balanced mode is where the reference capacitor is varied to give zero voltage on the output. The unbalanced mode requires measuring the ratio of output to excitation voltage to determine pressure.

This type of pressure measurement is quite accurate and has a wide operating range. Capacitive pressure measurement is also quite common for determining the level in a tank or vessel.

Transducer advantages

- Inaccuracy 0.01–0.2%
- Range of 80 Pa–35 MPa
- Linearity
- Fast response.

Transducer limitations

- Temperature sensitive
- Stray capacitance problems
- Vibration
- Limited over pressure capability
- Cost.

Many of the disadvantages above have been addressed and their problems reduced in newer designs.

Temperature-controlled sensors are available for applications requiring a high accuracy. With strain gauges being the most popular form of pressure measurement, capacitance sensors are the next most common solution.

2.8.5   Linear variable differential transformer

This type of pressure measurement relies on the movement of a high permeability core within transformer coils. The movement is transferred from the process medium to the core by use of a diaphragm, bellows or bourdon tube.

The LVDT operates on the inductance ratio between the coils. Three coils are wound onto the same insulating tube containing the high permeability iron core. The primary coil is located between the two secondary coils and is energized with an alternating current.

Equal voltages are induced in the secondary coils if the core is in the center. The voltages are induced by the magnetic flux. When the core is moved from the center position, the result of the voltages in the secondary windings will be different. The secondary coils are usually wired in series (Figure 2.11).

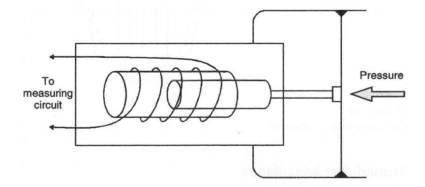

Figure 2.11
Schematic representation of a linear motion variable inductance prior transducer element (LMVIPTE)

Transducer limitations or disabilities

- Mechanical wear
- Vibration.

Summary

- This is an older technology, used before strain gauges were developed.
- Typically found on old weigh-frames or may be used for position control applications.
- Very seldom used anymore; strain gauge types have superseded these transducers in most applications.

2.8.6   Optical

Optical sensors can be used to measure the movement of a diaphragm due to pressure. An opaque vane is mounted onto a diaphragm and moves in front of an infrared light beam. As the light is disturbed, the received light on the measuring diode indicates the position of the diaphragm.

A reference diode is used to compensate for the aging of the light source. Also, by using a reference diode, the temperature effects are canceled as they affect the sensing and reference diodes in the same way (Figure 2.12).

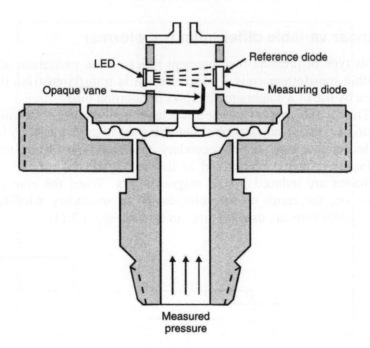

Figure 2.12
Optical pressure transducer

Transducer advantages

- Temperature corrected
- Good repeatability

- Negligible hysteresis as optical sensors require very little movement for accurate sensing
- Optical pressure measurement provides very good repeatability with negligible hysteresis.

Transducer limitations or disabilities

- Expensive.

2.8.7 Pressure measurement applications

There are a number of requirements that need to be considered with applications in pressure measurement. Some of the more important of these are listed below:

- Location of process connections
- Isolation valves
- Use of impulse tubing
- Test and drain valves
- Sensor construction
- Temperature effects
- Remote diaphragm seals
- Corrosion may cause a problem to the transmitter and pressure sensing element
- The sensing fluid contains suspended solids or is sufficiently viscous to clog the piping
- The process temperature is outside of the normal operating range of the transmitter
- The process fluid may freeze or solidify in the transmitter or impulse piping
- The process medium needs to be flushed out of the process connections when changing batches
- Maintaining sanitary or aseptic conditions
- Eliminating the maintenance required with wet leg applications
- Making density or other measurements.

2.9 Flow meters

In many industrial applications it is convenient and useful to measure flow and so a large percentage of transmitter sales are for measuring flow. As a result, there is a huge range of flowmeters to suit a variety of applications. The operation of these may conform to one of two approaches.

2.9.1 Energy-extractive flowmeters

This is the older of the two approaches, and uses flow measurement devices that reduce the energy of the system. The most common of these are the differential pressure producing flowmeters, such as the orifice plate, flow nozzle and venturi tube (Figure 2.13).

Orifice plate

A standard orifice plate is simply a smooth disk with a round, sharp-edged inflow aperture and mounting rings. In the case of viscous liquids, the upstream edge of the bore can be rounded. The shape of the opening and its location do vary widely, and this is dependent on the material being measured. Most common are concentric orifice plates with a round opening in the center. They produce the best results in turbulent flows when used with clean liquids and gases.

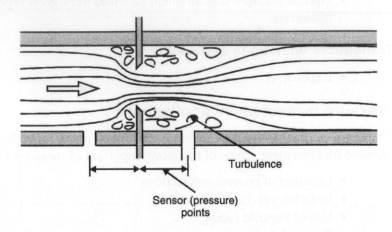

Figure 2.13
Flow patterns with an orifice plate

When measuring liquids the bore can be positioned at the top of the pipeline to allow the passage of gases. The same applies when allowing suspended solids to pass by positioning the bore at the bottom and gaining a more accurate liquid flow measurement.

Standard orifice meters are primarily used to measure gas and vapor flow. Measurement is relatively accurate; however because of the obstruction of flow there is a relatively high residual permanent pressure loss. They are well-understood, rugged and relatively inexpensive for large pipe sizes and are suited for most clean fluids and aren't influenced by high temperatures.

Transducer advantages

- Simple construction
- Inexpensive
- Easily fitted between flanges
- No moving parts
- Large range of sizes and opening ratios
- Suitable for most gases and liquids
- Well understood and proven
- Price does not increase dramatically with size.

Transducer limitations or disabilities

- Inaccuracy, typically 1%
- Low rangeability, typically 4:1
- Accuracy is affected by density, pressure and viscosity fluctuations
- Erosion and physical damage to the restriction affects measurement accuracy
- Cause some unrecoverable pressure loss
- Viscosity limits measuring range
- Require straight pipe runs to ensure if accuracy is maintained
- Pipeline must be full (typically for liquids)

- The inaccuracy with orifice-type measurement is due mainly to process conditions and temperature and pressure variations
- They are also affected by ambient conditions and upstream and downstream piping, as this affects the pressure and continuity of flow.

Turbine or rotor flow transducer

Turbine meters have rotor-mounted blades that rotate when a fluid pushes against them. They work on the reverse concept to a propeller system. In a propeller system, the propeller drives the flow; in this case the flow drives and rotates the propeller. Since it is no longer propelling the fluid, it's now called a turbine. The rotational speed of the turbine is proportional to the velocity of the fluid.

Different methods are used to convey rotational speed information. The usual method is by electrical means where a magnetic pick-up or inductive proximity switch detects the rotor blades as they turn. As each blade tip on the rotor passes the coil, it changes the flux and produces a pulse. The rate of pulses indicates the flow rate through the pipe.

Turbine meters require a good laminar flow. In fact 10 pipe diameters of straight line upstream, and no less than 5 pipe diameters downstream from the meter are required. They are therefore not accurate with swirling flows.

Turbine meters are specified with minimum and maximum linear flow rates that ensure the response is linear and the other specifications are met. For good rangeability, it is recommended that the meter be sized such that the maximum flow rate of the application be about 70–80% of that of the meter. Density changes have little effect on the meter's calibration.

Transducer advantages

- High accuracy, repeatability and rangeability for a defined viscosity and measuring range
- Temperature range of fluid measurement: –220 to +350 °C
- Very high pressure capability: 9300 psi
- Measurement of non-conductive liquids
- Capability of heating measuring device
- Suitable for very low flow rates.

Transducer limitations or disabilities

- Not suitable for high viscous fluids
- Viscosity must be known
- 10D upstream and 5D downstream of straight pipe is required
- Not effective with swirling fluids
- Only suitable for clean liquids and gases
- Pipe system must not vibrate
- Specifications critical for measuring range and viscosity
- As turbine meters rely on the flow, they do absorb some pressure from the flow to propel the turbine
- The pressure drop is typically around 20–30 kPa at the maximum flow rate and does vary depending on flow rate
- It is a requirement in operating turbine meters that sufficient line pressure be maintained to prevent liquid cavitation

- The minimum pressure occurs at the rotor; however the pressure recovers substantially at the turbine
- If the back-pressure is not sufficient, then it should be increased or a larger meter chosen to operate in a lower operating range – this does have the limitation of reducing the meter flow range and accuracy.

Summary

Turbine meters provide excellent accuracy, repeatability and rangeability for a defined viscosity and measuring range, and are commonly used for custody transfer applications of clean liquids and gases.

2.9.2   Energy-additive flow meters

A common example of the energy-additive approach is the magnetic flowmeter, illustrated in Figure 2.14. This device is used to make flow measurements on a conductive liquid. A charged particle moving through the magnetic field produces a voltage proportional to the velocity of the particle. A conductive liquid consisting of charged particles will then produce a voltage proportional to the volumetric flow rate.

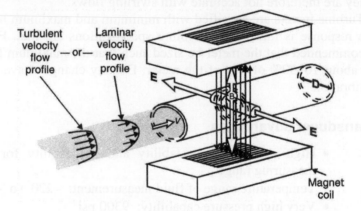

Figure 2.14
Schematic representation of a magnetic flowmeter

The magmeter

The advantages of magnetic flowmeters are:

- They have no obstructions or restrictions to flow
- No pressure drop or differential
- No moving parts to wear out
- They can accommodate solids in suspension
- No pressure sensing points to block up
- They measure volume rate at the flowing temperature independent of the effects of viscosity, density, pressure or turbulence
- Another advantage is that many magmeters are capable of measuring flow in either direction.

Most industrial liquids can be measured by magnetic flowmeters; these include acids, bases, water and aqueous solutions. However some exceptions are most organic chemicals and refinery products which have insufficient conductivity for measurement. Also pure substances, hydrocarbons and gases cannot be measured.

In general the pipeline must be full, although with newer models, level sensing takes this factor into account when calculating a flow rate.

Accuracy

Magnetic flowmeters are very accurate and have a linear relationship between the output and flow rate. Alternatively, the flow rate can be transmitted as a pulse per unit of volume or time.

The accuracy of most magnetic flowmeter systems is 1% of full-scale measurement. This takes into account both the meter itself and the secondary instrument. Because of its linearity, the accuracy of low flow rates exceeds that of such devices as the Venturi tube. The magnetic flowmeter can be calibrated to an accuracy of 0.5% of full scale and is linear throughout.

Selection, sizing and liners

Sizing of magmeters is done from manufacturer's nomographs to determine suitable diameter meters for flow rates.

The principle of operation of the magmeter requires the generation of a magnetic field and the detection of the voltage across the flow.

If the pipe is made of a material with magnetic properties, then this will disrupt the magnetic field and effectively short circuit the magnetic field. Likewise if the inside of the pipe is conductive, then this will short circuit the electrodes used to detect the voltage across the flow.

The meter piping must be manufactured from a non-magnetic material such as stainless steel in order to prevent short circuiting of the magnetic field.

The lining of the meter piping must also be lined with an insulating material to prevent short circuiting of the electric field.

The liner has to be chosen to suit the application, particularly the resistance it has to the following:

- Chemical corrosion
- Erosion
- Abrasion
- Pressure
- Temperature.

Liner materials

Teflon (Polytetrafluoroethylene (PTFE) resin)

- Widely used due to its high temperature rating
- Anti-stick properties reduce problems with build-up
- Approved for food and beverage environments
- Resistant to many acids and bases.

Neoprene

- Good abrasion resistance
- Good chemical resistance.

Soft rubber

- Relatively inexpensive
- High resistance to abrasion
- Used mainly for slurry applications.

Hard rubber

- Inexpensive
- General-purpose applications
- Used mainly for water and soft slurries.

Ceramic

- High abrasion resistance
- High corrosion resistance
- High temperature rating
- Less expensive to manufacture
- Also suited to sanitary applications
- Strong compressive strength, but poor tensile strength
- Brittle
- May crack with sudden temperature changes, especially downward
- Cannot be used with oxidizing acids or hot concentrated caustic.

Installation techniques

For correct operation of the magmeter, the pipeline must be full. This is generally done by maintaining sufficient back-pressure from downstream piping and equipment. Meters are available that make allowance for this problem, but are more expensive and are specialized. This is mainly a problem in gravity feed systems.

Magmeters are not greatly affected by the profile of the flow and are not affected by viscosity or the consistency of the liquid. It is however recommended that the meter be installed with 5 diameters of straight pipe upstream and 3 diameters of straight pipe downstream from the meter.

Applications requiring reduction in the pipe diameter for the meter installation need to allow for the extra length of reducing pipe. It is also recommended that in those applications, the reducing angle not be greater than 8°, although manufacturer's data should be sought.

Grounding is another important aspect when installing magmeters, and manufacturer's recommendations should be adhered to. Such recommendations would require the use of copper braid between the meter flange and pipe flange at both ends of the meter. These connections provide a path for stray currents and should also be grounded to a suitable grounding point. Magmeters with built-in grounding electrodes eliminate this problem, as the grounding electrode is connected to the supply ground.

Transducer advantages

- No restrictions to flow
- No pressure loss
- No moving parts
- Good resistance to erosion
- Independent of viscosity, density, pressure and turbulence

- Good accuracy
- Bi-directional
- Large range of flow rates and diameters.

Transducer limitations or disabilities

- Expensive
- Most require a full pipeline
- Limited to conductive liquids
- As mentioned earlier, a magnetic flowmeter consists of either a lined metal tube, usually stainless steel because of its magnetic properties, or an unlined non-metallic tube. The problem can arise if the insulating liners and electrodes of the magnetic flowmeter become coated with conductive residues deposited by the flowing fluid
- Erroneous voltages can be sensed if the lining becomes conductive
- Maintaining high flow rates reduces the chances of this happening. However, some manufacturers do provide magmeters with built-in electrode cleaners
- Block valves are used on either side of AC-type magmeters to produce zero flow and maintain full pipe to periodically check the zero. DC units do not have this requirement.

Ultrasonic flow measurement

There are two types of ultrasonic flow measurement:

1. Transit-time measurement, used for clean fluids
2. Doppler effect, used for dirty, slurry-type flows.

Transit-time ultrasonic flow measurement

The transit-time flowmeter device sends pulses of ultrasonic energy diagonally across the pipe. The transit time is measured from when the transmitter sends the pulse to when the receiver detects the pulse.

Each location contains a transmitter and receiver. The pulses are sent alternatively upstream and downstream and the velocity of the flow is calculated from the time difference between the two directions.

Transit-time ultrasonic flow measurement is suited for clean fluids. Some of the more common process fluids consist of water, liquefied gases and natural gas.

Doppler effect ultrasonic flow measurement

The Doppler effect device relies on objects with varying density in the flow stream to return the ultrasonic energy. With the Doppler effect meter a beam of ultrasonic energy is transmitted diagonally through the pipe. Portions of this ultrasonic energy are reflected back from particles in the stream of varying density. Since the objects are moving, the reflected ultrasonic energy will have a different frequency. The amount of difference between the original and returned signals is proportional to the flow velocity.

General summary

Most ultrasonic flowmeters are mounted on the outside of the pipe and as such operate without coming in contact with the fluid. Apart from not obstructing the flow, they are not affected by corrosion, erosion or viscosity. Most ultrasonic flowmeters are bi-directional and sense flow in either direction.

Advantages

- Suitable for large diameter pipes
- No obstructions, no pressure loss
- No moving parts, long operating life
- Fast response
- Installed on existing installations
- Not affected by fluid properties.

Transducer limitations or disabilities

- Accuracy is dependant on flow profile
- Fluid must be acoustically transparent
- Errors caused by build-up in pipe
- Only possible in limited applications
- Expensive
- Pipeline must be full
- Turbulence or even the swirling of the process fluid can affect the ultrasonic signals
- In typical applications the flow needs to be stable to achieve good flow measurement, and typically this is done by allowing sufficient straight pipe up and downstream of the transducers
- The straight section of pipe upstream would need to be 10–20 pipe diameters with the downstream requirement of 5 pipe diameters
- For the transit time meter, the ultrasonic signal is required to traverse across the flow, therefore the liquid must be relatively free of solids and air bubbles
- Anything of a different density (higher or lower) from the process fluid will affect the ultrasonic signal.

Summary

Doppler flowmeters are not high accuracy or high performance devices, but do offer an inexpensive form of flow monitoring. Their intended operation is for dirty fluids, and find applications in sewage, sludge and waste water processes.

Being dependent on sound characteristics, ultrasonic devices are dependent on the flow profile and are also affected by temperature and density changes.

2.10   Level transmitters

There are numerous ways to measure level that require differing technologies and to encompass all the various units of measurement.

- Ultrasonic, transit time
- Pulse echo
- Pulse radar
- Pressure, hydrostatic

- Weight, strain gauge
- Conductivity
- Capacitive.

For continuous measurement, the level is detected and converted into a signal that is proportional to the level. Microprocessor-based devices can indicate level or volume.

Different techniques also have different requirements. For example, when detecting the level from the top of a tank, the shape of the tank is required to deduce volume.

When using hydrostatic means, which detects the pressure from the bottom of the tank, the density is to be known and remains constant. Level sensing is a simpler concept than most other process variables and allows a very simple form of control. The sensors can be roughly grouped into categories according to the primary level sensing principle involved.

The signals produced by these means must then be converted into a signal suitable for process control applications, such as an electrical, pneumatic or digital signal.

2.10.1   Installation considerations

The following are outlines of the more important considerations that need to be considered when installing either atmospheric or pressurized vessels.

Atmospheric vessels

Most instruments involved with level detection can be easily removed from the vessel. Top mounting of the sensing device also eliminates the possibility of process fluid entering the transducer or sensor housing should be the nozzle or probe corrode or breaks off. Many level measurement devices have the added advantage that they can be manually gauged. This provides two important factors:

1. Measurements are still possible in the event of equipment failure
2. Calibration and point checks can provide vital operational information.

One common installation criteria for point detection devices is that they be mounted at the actuation level, which may present accessibility problems.

Pressurized vessels

Two main considerations apply with level measurement devices in pressurized vessels:

1. Facilities for removal and installation while the vessel is pressurized
2. The pressure rating of the equipment for the service.

Pressurized vessels can also be used to prevent fugitive emissions, where an inert gas such as hydrogen can be used to pressurize the process materials. Compensation within the level device needs also to be accounted for as the head pressure changes.

The accuracy of the measuring device may be dependent on the following:

- Gravity variations
- Temperature effects
- Dielectric constant.

Also the presence of foam, vapor or accumulated scum on the transducer affects the performance.

Impact on the overall control loop

Level sensing equipment is generally fast responding, and in terms of automated continuous control, does not add much of a lag to the system.

It is good practice though, to include any high and low switch limits into the control system. If the instrumentation does fail or goes out of calibration, then the process information can be acquired from the high and low limits. Apart from the hard-wired safety circuits, it is good practice to incorporate this information into the control system.

Future technologies

The cost of sensing equipment is not a major consideration compared with the economics of controlling the process. There is therefore a growing demand for accuracy in level measuring equipment.

Newer models incorporate better means of compensation, but not necessarily new technologies. Incorporating a temperature compensation detector in the pressure-sensing diaphragm provides compensation and also an alternative to remote pressure seals and ensures the accuracy and stability of the measurement.

Greater demands in plant efficiency may require an improved accuracy of a device, not just for the actual measurement, but also to increase the range of operation. If the safety limits were set at 90% due to inaccuracies with the sensing device, then an increased range could be achieved by using more accurate equipment.

Demands are also imposed on processes to conform to environmental regulations. Accurate accounting of materials assist in achieving this. Such technologies as RF admittance or ultrasonic minimize the expense of this environmental compliance.

Problems occur in trying to sense level in existing vessels that may be non-metallic. RF flexible cable sensors have an integral ground element which eliminates the need for an external ground reference when using the sensor to measure the level of process materials in non-metallic vessels.

2.11   The spectrum of user models in measuring transducers

As an example of the extremes that can occur between the same type of measuring transducer, consider the case of the thermocouple. Firstly in its most simple form, it consists of two dissimilar metal wires joined together to form a loop consisting of two junctions or connections. The Seebeck effect (If the temperature of the two junctions is different, a current will flow in the loop.) then comes into play. Looked at in practice, a thermocouple-measuring circuit actually measures the *difference* between the two junctions forming the circuit.

Unfortunately three major problems occur with this form of temperature measurement.

2.11.1   Voltage generation of a thermocouple

Only very low emfs are generated, typically around 1.8 to 6.0×10^{-12} V per 0 °C; so electrically induced noise, either as the normal or common mode type, can become a problem. Normal mode noise being the more difficult one to remove from a system, this usually being achieved by the introduction of guard lead wires, or careful cable screening.

2.11.2   Thermocouple linearity

The output of any type of thermocouple is not linear relative to the applied or measured temperature range, This can cause linearity, scaling, ranging and calibration problems.

2.11.3   Cold junction compensation

To complete any electrical circuit, requires the formation of a loop, so in the case of the thermocouple, as second junction has to exist to achieve this. This is called the 'cold junction' and usually sits at ambient temperature, which of course varies and introduces measurement errors which can be extremely large, especially if measurement of the physical quantity is close to the ambient or cold junction temperature.

The simplest form of thermocouple application is in the form of a galvanometer which has the sensitivity to measure the low voltages involved. This is equipped with a temperature-sensitive compensating resistor, located next to the input terminals where the measuring circuit's cold junction is. This resistor forms part of that measuring circuit and corrects the effect, by changing its resistance and hence the current flow in the circuit, of ambient temperature changes.

The problem with this arrangement is that it is direct reading, and hence does not easily lend itself to inclusion in process control systems, and the physical circuit from the indicator to the thermocouple measuring tip has to be 'tuned' to a specific resistance for the cold junction compensator to be accurate. To overcome the non-linearity problem, the scale of the instrument is scaled to the 'profile' of the related response curve of the thermocouple type being used.

2.12   Instrumentation and transducer considerations

There are many considerations that have to be taken into account when selecting instruments and transducers. The following is an explanation and index of the more important aspects of choice.

2.12.1   Signal transmission pneumatic vs electronic

Electronic means for signal transmission and control is becoming more favored, however pneumatic controls are still used and do have advantages in different applications.

Advantages – electronic

- Lower installation cost
- Lower maintenance
- Higher accuracy (especially smart instruments)
- Faster dynamic response
- Suitable for long distances
- Digital control system compatible.

The primary reason for selecting electronic devices is their compatibility with the control system. With data exchange highways becoming more common it is also easier to obtain more information from the sensor with smart electronics.

Advantages – pneumatic

- Lower initial hardware cost
- Simple design
- Less affected by corrosive environments

- Easily connected with control valves
- Pneumatics has a prime advantage because of their safety in hazardous locations.

2.12.2   Signal conditioners

Signal conditioners change or alter signals so that different process devices can effectively and accurately communicate with each other. They are typically used to link process instruments with indicators, recorders, and microprocessor-based control and monitoring systems.

They consist of either:

Signal conversion   A signal converter is used to change an analog signal from one form to another. This enables equipment with differing signals to communicate.

Signal boosting   For analog signals (voltage) that are required to be transmitted over long distances, it is possible that the signal may attenuate, or fade. For analog signals (current) in loops that have a number of loop-powered devices, the signal may not be strong enough.

2.12.3   Noise

Electrical noise, or interference, is unwanted electrical signals that cause disruptive errors, or even completely disable electronic control and measuring equipment.

There are two main categories of electrical measurement noise:

1. Radio frequency interference (RFI)
2. Electromagnetic interference (EMI).

Some examples of the more commonly encountered sources of interference are:

- Hand-held (walkie-talkie)
- Cellular phones
- AC and DC motors
- Transformers
- Arc welders
- Large solenoids, contactors and relays
- High power cabling, both voltage and current
- High speed power switching, such as SCRs and thyristor
- Variable frequency drives
- Static discharges
- Induction heating systems
- Radar devices
- Fluorescent lights.

Radio frequency interference and electromagnetic interference can cause unpredictable performance in instrumentation. These types of interference can often be non-repeatable, making it hard to detect, isolate and rectify the problem. RFI and EMI can also degrade an instrument's performance and possibly cause the instrument to fail completely.

Any of these problems can result in reduced production rates, process inefficiency, plant shutdowns and possibly even create dangerous safety hazards.

There are two basic approaches to protecting instrumentation systems from the harmful effects of RFI and EMI.

1. The first is to keep the interference from entering the system by:

   - Shielding
   - Proper grounding
   - Terminal filters.

2. The second is to design the system so that it is unaffected by RFI and EMI.

Noise reduction techniques

Some of the more common techniques for reducing or even eliminating electrically induced noise are:

- *Use of transmitters, i.e. for thermocouples*:   The signal is more robust to noise over long distances. Typically 4–20 mA.
- *Shielded/twisted pair cable*:   Twisting is done to decouple the wires from induced currents from varying electric and magnetic fields that may exist. The principle of twisting is that equal voltages are induced in each loop of the twisted wires, but of opposite phase which makes them cancel.
- *AC-inductive load circuits*:   For AC-inductive loads, use a properly rated MOV across the load in parallel with a series RC snubber. An effective RC snubber circuit would consist of a 0.1 μF capacitor of suitable voltage rating, and a 47 Ω 0.5 W resistor.
- *DC-Inductive load circuits*:   For DC-inductive loads, use of a diode across the load is effective, provided the polarity is correct. Use of an RC snubber circuit can be added as an enhancement.

2.12.4   Materials of construction

Often when selecting measurement or control equipment, options are available for the various materials of construction. The primary concern is that the process material will not cause deterioration or damage to the device.

Below is a brief list of other qualities or characteristics that assist in the selection of the material of construction.

- 316SS
- Hastelloy C-276
- Monel
- Carbon steel
- Beryllium copper; good elastic qualities
- Ni-Span C; very low temperature coefficient of elasticity
- Inconel; extreme operating temperatures and corrosive process
- Stainless steel; extreme operating temperatures and corrosive process
- Quartz; minimum hysteresis and drift.

2.12.5   Signal linearization

When the output of a device responds at a proportional rate to changes in the input, then the device is linear and there is a constant gain (output / input) over the full range of operation and the resolution remains constant. If the response or reaction of some device in a system is not linear then it may need to be made linear because there are two main problems, when the device is not linear:

1. The gain changes
2. The resolution and accuracy change.

In a control system there are three ways to account for non-linear equipment:

1. Base application on the highest gain
2. Measure the gain at a number of points
3. Modify the gain as a function of the process variable.

The simpler way to overcome any non-linearity is to linearize the signal before the control system calculations.

2.13   Selection criteria and considerations

Reasons for selecting one type of measuring equipment over another vary, but typically the decisions are based on the perceived advantages and disadvantages of the range of devices available.

A comprehensive list would take into account the following:

- Accuracy
- Reliability
- Purchase price
- Installed cost
- Cost of ownership
- Ease of use
- Process medium, liquid/stem/gas
- Degree of smartness
- Repeatability
- Intrusiveness
- Sizes available
- Maintenance
- Sensitivity to vibration.

In addition particular requirements for flow would include:

- Capability of measuring liquid, steam and gas
- Rangeability
- Turndown
- Pressure drop
- Reynolds number
- Up and downstream piping requirements.

A more systematic approach to selection process measurement equipment would cover the following steps.

2.13.1   Application

These are the requirement and purpose of the measurement.

- Monitor
- Control

- Indicate
- Point or continuous
- Alarm.

2.13.2 Processed material properties

Many process-measuring devices are limited by the process material that they can measure.

- Solids, liquids, gas or steam
- Conductivity
- Multi-phase, liquid/gas ratio
- Viscosity
- Pressure
- Temperature.

2.13.3 Performance

This relates to the performance required in the application.

- Range of operation
- Accuracy
- Linearity (accuracy may include linearity effects)
- Repeatability (accuracy may include repeatability effects)
- Response time.

2.13.4 Installation

Mounting is one of the main concerns, but the installation does involve the access and other environmental concerns.

- Mounting
- Line size
- Vibration
- Access
- Submergence.

2.13.5 Economics

The associated costs determine whether the device is within the budget for the application.

- Purchase cost
- Installation cost
- Maintenance cost
- Reliability/replacement cost.

2.13.6   Environment and safety

This relates to the performance of the equipment to maintain the operational specifications, and also failure and redundancy should be considered.

- Process emissions
- Hazardous waste disposal
- Leak potential
- Trigger system shutdown.

2.13.7   Measuring devices and technology

At this stage the selection criteria is established and weighed up with readily available equipment. A typical example for flow is shown:

1. DP

   - Orifice plate
   - Variable area

2. Velocity

   - Magmeter
   - Vortex
   - Turbine
   - Propeller

3. Positive displacement

   - Oval gear
   - Rotary

4. Mass flow

   - Coriolis
   - Thermal.

2.13.8   Vendor supply

Limitations may be imposed, particularly with larger companies that have preferred suppliers, in which case the selections may be limited, or the procedure for purchasing new equipment may not warrant the time and effort for the application.

2.14   Introduction to the smart transmitter

The most elaborate form of thermocouple transducer, quite often referred to as a '*smart transmitter*' as shown in Figure 2.15, comprises:

- An electronic cold junction compensator
- A highly stable DC amplifier to get the thermocouples low voltages up to a reasonable operating level
- Some form of microprocessor that perform a linearization function on the thermocouple's generated voltage
- A ranging function

- An output or transmitter part of the system, where selectable types of output can be selected
- And, of course, the thermocouple itself.

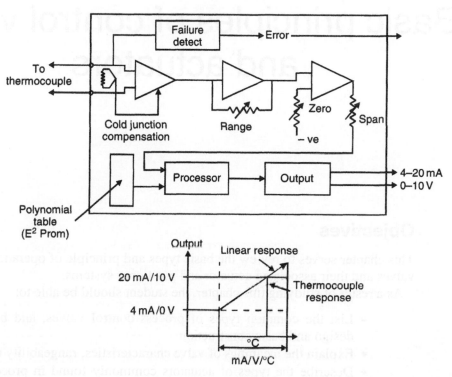

Figure 2.15
Microprocessor-based thermocouple measuring system with ranged and linear output

This then gives the availability to use a thermocouple of a type with a total range of 0–650 °C, to be ranged, or set for 100–300 °C and this, with a 4–20 mA transmitter output, gives an output = 12.5 °C/mA, linearly through the required or selected range.

In normal use, using the entire range of the thermocouple, we would have an output sensitivity of 650 °C/20–4 mA or 40.625 °C/mA giving a sensitivity ratio of 3.25:1

This concept can be applied to most types of measurement transducers, conceptually saying that the output of these devices can be ranged and made linear before being introduced to the controllers inpsut itself.

3

Basic principles of control valves and actuators

3.1 Objectives

This chapter serves to review the basic types and principle of operation of process control valves and their associated actuator and positioner systems.

As a result of studying this chapter, the student should be able to:

- List the common types of process control valves, and briefly describe their design and basic construction
- Explain the meanings of valve characteristics, rangeability and sizing
- Describe the types of actuators commonly found in process control systems, and list their applications.

3.2 An overview of eight of the most basic types of control valves

In most process control systems the final control element, driven by the output of the process controller, is usually some form of valve. This chapter serves to introduce the student to eight of the most common types of control valves, flow throttling devices and the basic range of actuators used to control them.

Sections 3.2.1 through to 3.2.8 describe the various types of valves in question, starting with an overview and general description, the types and variances within their manufactured ranges, sizes, design pressures and temperature ranges and their rangeability.

Any special attributes or uses a valve may have are also described.

Section 3.2.9 introduces the reader to some of the more unusual types of valves, their design and usages.

3.2.1 Ball valves

Overview

The rotary ball valve, which used to be considered as an on–off shut-off valve is now used quite extensively as a flow control device. Some of the advantages include lower cost and weight, high flow capacity, tight shut-off and fire-safe designs. The ball valve contains a spherical plug that controls the flow of fluid through the valve body. Ball and cage valves are close to linear in terms of percent of flow or C_V to percent of stem or ball rotation. The three basic types of ball valve are listed below.

Types of ball valves

- *Conventional*:   1/4 turn pierced ball type (Figure 3.1)
- *Characterized*:   V and U notched along with a parabolic ball type (Figure 3.2)
- *Cage*:   Positioning a ball by means of a cage in relation to a seat ring and discharge port is used for control.

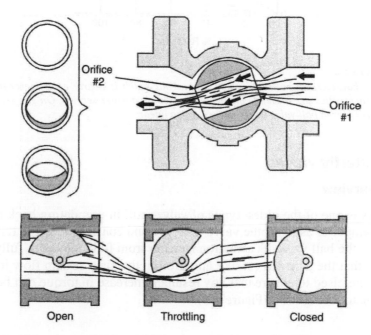

Figure 3.1
Cross-sectional views of conventional and characterized ball valves

Size and design pressure

- 0.5–42 in. (12.5 mm–1.06 m) in ANSI class 150 to 12 in. (300 mm) in ANSI class 2500
- Segmented ball – 1–24 in. (50–600 mm) in ANSI class 150–16 in. (400 mm) in ANSI class 300
- Pressure up to 2500 psig (17 MPa).

Design temperature

- Varies with design and material but typically –160 to +310 °C
- Special designs extend this range from − 180 to > +1000 °C.

Rangeability

Generally claimed to be about 50:1. Refer to Section 3.3.3.

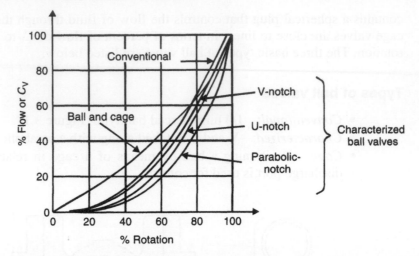

Figure 3.2
The characterized ball valve with a parabolic-notch is nearly equal percentage, while the ball and cage characteristics are closer to linear, when used on a water service. On gas services at critical velocities, the characterized ball valve lines move closer to linear

3.2.2   Butterfly valves

Overview

This is one of the oldest types of valves still in use, dating back from the 1920s. It acts as a damper or as a throttle valve in a pipe and consists of a disk turning on a diametral axis. Like the ball valve its actuation rotation from fully closed to fully open is 90°. Due to the fact that the disk can act like an airfoil in the main stream flow it is controlling, care must be exercised to ensure that any resultant increase in torque can be absorbed by the control actuator being used (Figure 3.3).

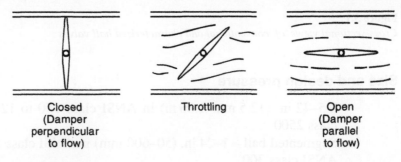

Closed	Throttling	Open
(Damper		(Damper
perpendicular		parallel
to flow)		to flow)

Figure 3.3
Vane positions of butterfly valve

Types of butterfly valves

- General purpose, aligned shaft, where the vane, disk, louver or flapper is rotated via the shaft to which it is attached.
- High-performance butterfly valve (HPBV), offset (eccentric) shaft, This design combines tight shut-off, reduced operating torque, and good throttling capabilities of the swing-through special disk shapes.

Size and design pressure

- To 48 in. (51 mm–1.22 m) are typical
- Units have been made from 0.75 to 200 in. (19 mm–5 m).

Design temperature

- Typically −260 to +540 °C
- Special designs extend this range up to > +1200 °C.

Rangeability

Generally claimed to be about 50:1 (Figure 3.4). Refer to Section 3.3.4.

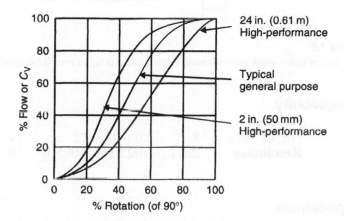

Figure 3.4
The flow characteristics of butterfly valves are affected by the position of the shaft (aligned or eccentric) and the relative size of the shaft compared to the valve size. For throttling purposes the valve is usually limited to rotate from 0° to 60° positions

3.2.3   Digital valves

Overview

Digital valves comprise a group of valve elements, or ports, assembled into a common manifold (Figure 3.5). Each element has a binary relationship with its neighbor which means that starting with the smallest port, the next port is twice the size of the previous one. The main advantages of this type of valve are their high speed, high precision and practically unlimited rangeability. Their biggest disadvantage is their high cost.

Size

$^3/_4$–10 in. (19–250 mm) in both line and angle patterns.

Design temperature

Cryogenic to +670 °C.

Design pressure limits

Up to 10 000 psig (690 bars).

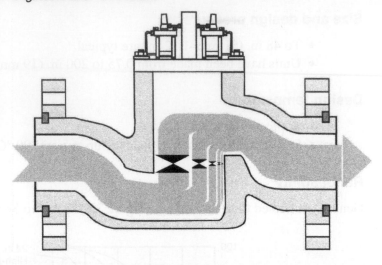

Figure 3.5
In a digital valve, each valve element is twice the size of its smaller neighbor

Rangeability

No. of 'bits'	8	10	12	14	16
Resolution	255:1	1023:1	4095:1	16, 383:1	65, 535:1

Applications

Where high speed (25–100 ms), accuracy, large rangeability and tight shut-off is needed.

3.2.4   Globe valves

Overview

Twenty or so years ago the majority of throttling control valves were of the globe type, characterized by linear plug movements and actuated by spring/diaphragm operators. The main advantages of the globe valve include the simplicity of the spring/diaphragm actuator, a wide range of characteristics, low cavitation and noise, a wide range of designs for corrosive, abrasive, high temperature and high pressure applications, a linear relationship between the control signal and valve stem movements and relative small amounts of dead band and hysteresis values (Figure 3.6).

Types of globe valves

- Single ported with characterized plug
- Single ported, cage guided
- Single ported, split body
- Double ported, top–bottom or skirt-guided plug
- Eccentric disk, rotary globe
- Angle
- Three way or Y type.

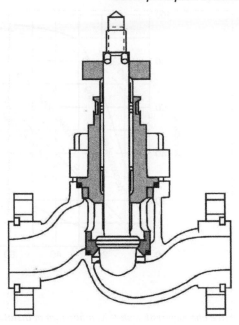

Figure 3.6
Cross section of a single ported globe valve

Flow characteristics

These characteristics are determined by the valve plug profile:

- Equal percentage
- Linear
- Quick opening.

Size and design pressure

- Generally $^1/_2$–14 in. (20 mm–356 m)
- Maximum size for type C is 6 in. (152 mm)
- Maximum size for type E is 12 in. (305 mm)
- Maximum size for type D is 16 in. (406 mm)
- Type F (the angle type) has been made in sizes up to 42 in. (1.05 m).

Typically all pressure ratings are available up to ANSI class 1500, with types B and D available through ANSI class 2500 and Types C and E are limited to ANSI class 600.

Design temperature

- Generally from − 200 to +540 °C
- Type B is limited to a maximum temperature of +400 °C
- Type C can operate down to − 260 °C.

Rangeability

If it is defined as the region within which the valve gain remains within 25% of the theoretical, it seldom exceeds 20:1 (Figure 3.7).

Manufacturers using other definitions claim 35:1. Refer to Section 3.3.3.

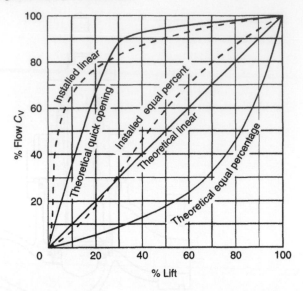

Figure 3.7
The theoretical valve characteristics shift as a function of installation. The dotted lines reflect such a shift in a mostly friction process where 100% flow, 20% pressure drop was assigned to the control valve

3.2.5   Pinch valves

Overview

These type of valves are called either pinch or clamp valves (Figure 3.8) depending on the configuration of the flexible tube and the means used for tube compression. They are also manufactured from a large range of materials such as teflon, PVC, neoprene, and polyurethane, each type of elastomer or plastic having its own particular application use.

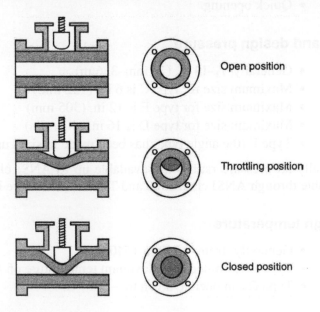

Open position

Throttling position

Closed position

Figure 3.8
Pinch valve with 'accurate closure'

This type of control valve, if carefully selected, has many advantages like high abrasion and corrosion resistance, packless construction, reasonable flow control rangeability, smooth flow, low replacement costs and a longer life than metal valves where abrasion and corrosion are present.

As in all things, though this valve has limitations such as pressure and temperature restraints due to the nature of the material used for the sleeves, and the number of operations, or flexing, that a particular type of liner can cope with, although a life span of >50 000 opening and closing cycles should be considered as the minimum.

Size

- 1–24 in. (25–610 mm)
- Special units from 0.1 to 72 in. (2.5 mm–1.8 m).

Design pressure

Generally up to ANSI class 150 with special units up to class 300.

Design temperature

Varies with design and material but typically –30 to +200 °C.

Rangeability

Generally claimed to be between 5:1 and 10:1. Refer to Section 3.3.3 (Figure 3.9).

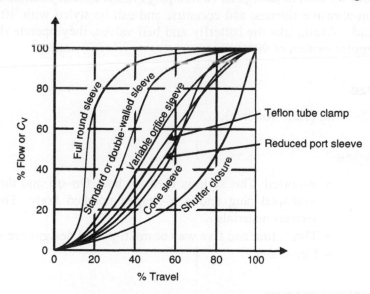

Figure 3.9
By reducing the port size or by making the sleeve 'cone' shaped, the characteristics are made more linear. The addition of a variable orifice within the sleeve provides flow throttling in the upper half of the stroke

3.2.6   Plug valves

Overview

Plug valves are probably the oldest type of valve in existence, being used in water distribution systems in ancient Rome and they probably pre-date the butterfly valve.

Consisting of a tapered vertical cylinder with a horizontal opening or flow-way inserted into the cavity of the valve body and due to the taper and lubricating system they use they are virtually leak proof to both gases and liquids (see Figure 3.10).

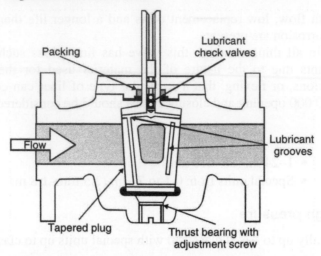

Figure 3.10
Typical sectional view of a plug valve

A very common use for this type of valve is in the tapping of beer barrels. They afford quick opening and closing action with tight leakproof closures under working pressures from vacuum to as high as 10 000 psig (70 MPa). They can be used for liquids, gases and non-abrasive slurries, and eccentric and can be styled with lift plugs for use with sticky fluids. Again, like the butterfly and ball valves, they operate through an actuator having angular motion of 90°.

Size

1/2–36 in. (12.5–960 mm).

Types

- V-ported: This style is used for both On–off and throttling control, utilizing a V-shaped plug and a V-shaped notched body. This is ideal for fibrous or viscous materials
- Three, four and five way or multi-ported designs are available
- Fire-sealed.

Design pressure

Typically from ANSI class 125 to ANSI class 300 ratings and up to 720 psig (5 MPa) pressure, with special units available for ANSI class 2500.

Design temperature

- Typically –70 to +200 °C
- Special units are available –160 to +315 °C.

Rangeability

Generally claimed to be between 20:1. Refer to Section 3.3.3 (Figure 3.11).

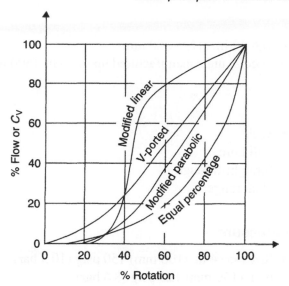

Figure 3.11
Plug valve characteristics are a function of the shape of the throttling plate or V-Port

3.2.7   Saunders diaphragm valves

Overview

The Saunders or diaphragm valve is sometimes also referred to as a weir valve (Figure 3.12). This valve operates by moving a flexible diaphragm toward or away from a weir. This valve can be considered as a half pinch valve as only one diaphragm is used, moving relative to a fixed weir; because of this however their flow characteristics are similar. The normal Saunders valve has a body with side section in the form of an inverted U shape, with the diaphragm closing the orifice at the top. A full-bore type is also available that has, when fully open, a fully rounded bore which is an important feature for ball-brush cleaning as required in applications like the food industry. It should be noted that mechanical damage can occur when opening this type of valve against a process vacuum.

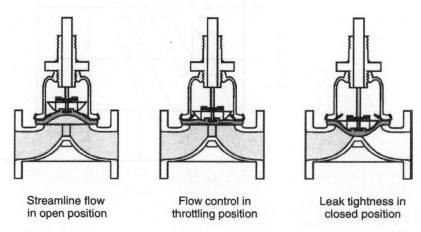

Streamline flow   Flow control in   Leak tightness in
in open position   throttling position   closed position

Figure 3.12
Main positions of weir-type Saunders control valve

Size

- 1/2–12 in. (12.5–300 mm)
- Special units manufactured up to 20 in. (500 mm).

Types

- Weir
- Full bore
- Straight-through
- Dual range.

Design pressure

- Sizes <= 4 in. (100 mm) 150 psig (10.3 bar)
- 6 in. (150 mm) 125 psig (8.6 bar)
- 8 in. (200 mm) 100 psig (6.9 bar)
- 10–12 in. (250–300 mm) 65 psig (4.5 bar).

Design temperature

- With most elastomer diaphragms –12 to +65 °C
- With PTFE diaphragms –34 to +175 °C.

Rangeability

Generally claimed to 10:1; refer to Section 3.3.3 (Figure 3.13).

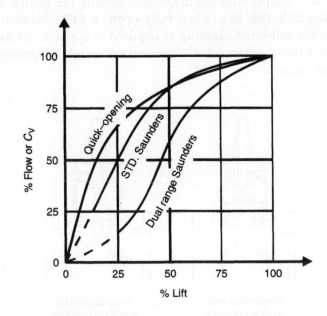

Figure 3.13
The characteristics of conventional Saunders valves are near to quick-opening, while the characteristics of the dual range design is closer to linear

3.2.8 Sliding gate valves

Overview

In this type of valve, the flow rate is controlled by sliding a plate with a hole in it past a stationary plate, usually placed at 90° to the line of flow, with a corresponding hole in it. These holes can be round, or shaped to profile the flow characteristic of the valve. This valve is sometimes used in automatic control but is not really considered a control valve. However, this type of valve can operate with pressures up to 10 000 psig (70 MPa). The accuracy of these valves, particularly in proportional control, depends solely on the accuracy of the chosen actuator (Figure 3.14).

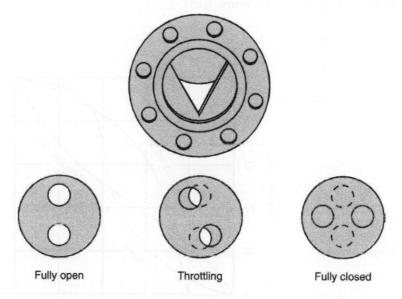

Fully open   Throttling   Fully closed

Figure 3.14
Typical sliding gate valve with 'V' Insert

Types

- Knife gate
- V-insert
- Plate and disk (multi-orifice)
- Positioned disk.

Size

- *On–off*: 2–120 in. (50 mm to 3.0 m)
- *Throttling*: $^{1}/_{2}$–24 in. (12–600 mm)
- *Throttling*: $^{1}/_{2}$–6 in. (12–150 mm)
- *Throttling*: 1 in. and 2 in. (25 mm and 50 mm).

Design pressure

- *Types A and B*:   Up to ANSI class 150
- *Type C*:   Up to ANSI class 300
- *Type D*:   Up to 10 000 psig (70 MPa).

Design temperature

- *Types A and B*:   Cryogenic to 260 °C
- *Type C and D*:   − 30 to +600 °C.

Rangeability

- *Types A*:   10:1
- *Type B*:   20:1
- *Type C*:   Up to 50:1 is claimed.

Refer to Section 3.3.3 (Figure 3.15).

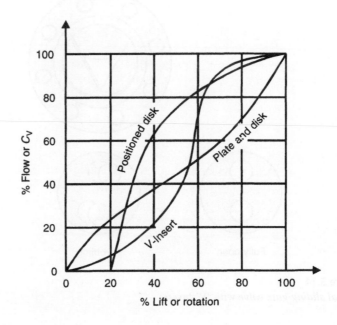

Figure 3.15
Characteristics of various sliding gate valve types

3.2.9   Special valve designs

This section is included to expose the student to some of the more uncommon types of valves that are currently being used. The reason for this is that with the current technical advancements and stringent requirements of accuracy, etc. that are now being made of process control systems, these valves are becoming more common. These valves are neither linear nor rotary in operation, but use other methods such as fluidics or static pressure of the process fluid in throttling the valve.

Dynamically balanced plug valves

This family of valves is used where there is no external power available to operate the valve, and therefore the static pressure of the process fluid is used to achieve throttling. The upstream, or back, pressure is used to move a plug against the force of a return spring. Variances in supply pressure affect the position of the plug relative to the spring tension. Control is achieved with a pilot valve poppet assembly (Figure 3.16).

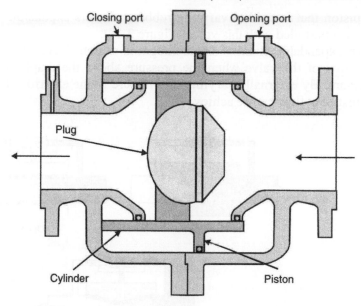

Figure 3.16
Sectional view showing the operation of the dynamically balanced plug valves

Diaphragm-operated cylinder in-line valves

This valve is used for high pressure gas services due to its low level of vibration, turbulence and noise. It consists of a low convolution diaphragm for positive sealing. Inlet to outlet pressures of 1400 and 600 MPa respectively are possible in the 2 in. (50 mm) size (Figure 3.17).

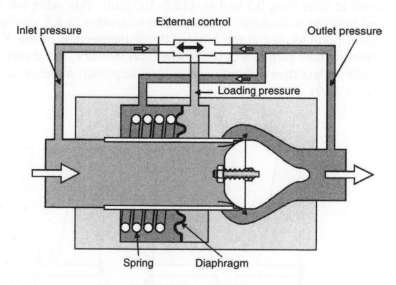

Figure 3.17
Sectional view of a diaphragm-operated cylinder in-line valve

Expandable element in-line valves

Streamlined flow of gas occurs in a valve where a solid rubber cylinder is expanded or contracted to change the area of an annular space. Control occurs via a hydraulic actuator or piston that is used to vary the rubber cylinders expansion. Pressures up to 1200 psid can be controlled with this valve (Figure 3.18).

An expandable element or diaphragm is stretched over a perforated dome shutting off the flow of the valve when the pressure above the diaphragm is greater than the line pressure. By externally varying the pressure to the exterior of the element control of the mainstream flow can be achieved.

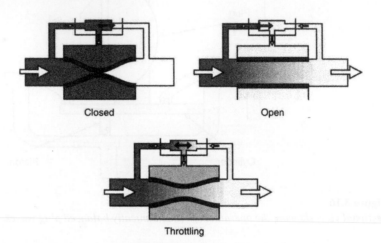

Figure 3.18
General view of an expandable element in-line valves

Fluid interaction valve

The *Coanda effect*, the basics of *Fluidics*, is used in this type of diverting valve which comes in sizes from 0.5 to 4 in. (12.5–100 mm). This valve has a flip-flop type of action used to divert a discharge from one port to another in a Y configuration by use of lateral control ports located at the base of the V intersection of the Y. This type of valve has numerous uses particularly in the chemical industry; the ability to divert a flow rapidly, usually in less than 100 ms, makes it an important member of the control valve family (Figure 3.19).

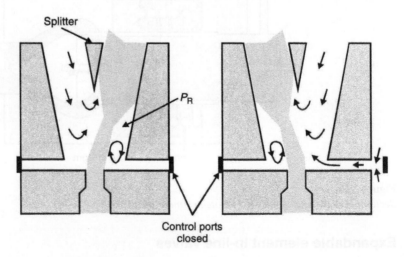

Figure 3.19
Operation of fluid interaction valves

3.3 Control valve gain, characteristics, distortion and rangeability

The characteristics, rangeabilities and gains of control valves are interrelated and a good understanding of these is necessary to be able to relate to the 'personality profile' of a process control valve.

3.3.1 Valve and loop gain

Gain is defined as $\Delta Output / \Delta Input$ and for a linear (constant gain) valve, valve gain (K_V) is defined as F_{max}/Stroke% or the maximum flow divided by the valve stroke in percentage.

The loop gain of a process control system (K_{LOOP}) should ideally be 0.5 to obtain quarter amplitude damping, an ideal and very stable state. Most process control loops consist of a minimum of four active units as listed below, each with their respective abbreviation indicated in [ ]'s as:

- A process control; P or proportional mode controller [K_C]
- The controller output driving a control valve [K_V]
- The valve effecting a process [K_P]
- A sensor/transducer measuring the process and feeding this as an input to the controller [K_S].

For the system to be stable all four components should have a linear gain and the overall product of their gains should equal 0.5 (for quarter damping).

$$K_{LOOP} = K_C \times K_V \times K_P \times K_S = 0.5$$

When a linear controller and sensor are used, and the gain of the process is also linear, a linear $(K_V = \text{constant})$ valve is needed to maintain the overall total loop gain constant at a value of 0.5.

However, if the process is non-linear (K_P varies with load), while the gains of K_C, K_V and K_S are constant, the value of K_{LOOP} will also vary about the optimum value of 0.5 resulting in either a sluggish or unstable operation of the process. The only way to maintain stability is for another component in the loop to change its gain in the opposite direction and magnitude to that of the process gain change. This can be either the controller gain (K_C) or the control valve gain (K_V). Here we will consider changes in the valve gain (K_V). When the control valve gain varies with its load (flow) it is named according to its characteristics (Figure 3.20), these being:

- *Equal percentage*: K_V increases at a *constant rate* with flow
- *Variable rate*: K_V increases according to the profile; *Parabolic, Hyperbolic*, etc.
- *Quick-opening*: K_V drops when the flow through the valve *increases*.

The theoretical valve gain invariably changes in actual use if the valve pressure differential varies with load, this is the case in most pumping systems where the valve differential drops with increasing flow rates thereby reducing the valve gain K_V. This tends to shift the gain of equal percentage valves toward that of the linear type.

In this case installing an equal percentage valve into the system often greatly assists in keeping the valve gain linear.

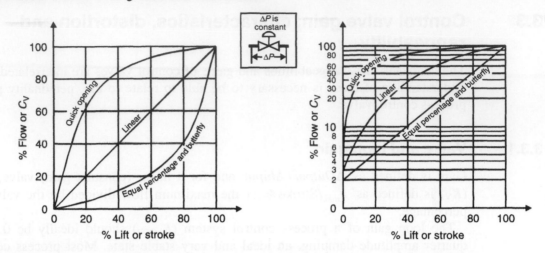

Figure 3.20
Inherent flow characteristics: quick opening, linear and equal percentage

The inherent characteristics of a control valve describes the relationship between the controller output as received by the actuator and the flow through the valve, assuming that:

- The actuator is linear (valve travel is proportional with controller output)
- The pressure difference ΔP across the valve is constant
- The process fluid is not flashing, subject to cavitation or at sonic velocity.

Selecting a valve characteristic can be a prolonged and complicated procedure; Driskell derived a general rule-of-thumb in selecting valve characteristics for the more common loops (Table 3.1):

Required Service	Valve ΔP < 2:1	Valve ΔP > 2:1 but < 5:1
Orifice type flow	Quick opening	Linear
Linear flow	Linear	Equal %
Level	Linear	Equal %
Gas pressure	Linear	Equal %
Liquid pressure	Equal %	Equal %

Table 3.1
List of common valve characteristics VV applications

In many cases the choice of valve characteristic has minimal effect on the loop parameters, and just about any type is acceptable for:

- Process with short time constants, such as flow control, most pressure control loops and temperature control when mixing a hot and cold stream
- Control loops operated by a narrow proportional band (high gain) controllers, such as most regulators
- Processes with a load variation of less than 2:1.

3.3.2 Valve distortion

Fluid flow through a valve is subjected to frictional losses, the consequence of this is shown in Figure 3.21. It can be seen from these curves that installation criteria can have substantial effects on a valve's flow characteristics and rangeability.

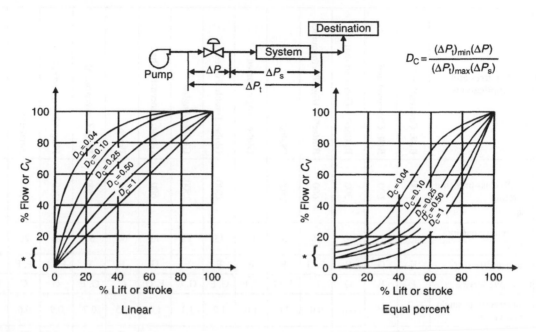

Figure 3.21
The effect of the distortion coefficient (D_C) on inherently linear and equal percentage valves, according to Boger

The *linear valve* has *a constant gain at all flow rates* and an *equal percentage* valve has a *gain directly proportional to flow.*

Therefore, if a loop tends toward oscillation at low flow rates (indicating a loop gain =>1) and is sluggish at high flow (indicating a gain <0.25) one should switch from a linear to equal percentage valve. The opposite therefore applies if oscillations occur at high flow rates, and a sluggish performance at low flow rates, change from an equal percentage to a linear control valve model.

3.3.3 Valve rangeability

Traditionally, rangeability has been defined as the ratio between minimum and maximum 'controllable' flow through a valve. The term 'minimum flow' (F_{MIN}) is defined as the flow below which the valve tends to close completely. Using this definition manufacturers usually claim:

- 50:1 rangeability for equal percentage valves
- 33:1 rangeability for linear valves
- 20:1 rangeability for quick opening valves.

This indicates that the flow through these valves can be controlled down to 2%, 3% and 5% respectively of their rated C_V. However it can be seen in Figure 3.21 that the minimum controllable flow rises as the distortion coefficient (D_C) drops.

Due to the fact that at minimum valve opening, the pressure drop through a valve, ΔP, is at a maximum, the valve will proportionally pass more flow.

Rangeability should be calculated as the ratio of the C_V required at maximum flow (minimum pressure drop) and the C_V required at minimum flow (maximum pressure drop).

Features and Applications	Ball: Conventional	Ball: Characterized	Butterfly: Conventional	Butterfly: High Performance	Digital:	Globe: Single Ported	Globe: Double Ported	Globe: Angled	Globe: Exocentric Disk	Pinch:	Plug: Conventional	Plug: Characterized	Sanders:	Sliding Gate V, Insert	Sliding Gate: Positioned Disk	Special Dynamically
ANSI class pressure rating	2500	600	300	600	2500	2500	2500	2500	600	150	2500	300	150	150	2500	1500
Maximum capacity (Cd)	45	25	40	25	14	12	15	12	13	60	35	25	20	30	10	30
Characteristics	F	G	P	F, G	E	E	E	E	G	P	P	F, G	P, F	F	F	F, G
Corrosive service	E	E	G	G	F, G	G, E	G, E	G, E	F, G	G	G, E	G	G	F, G	G	G, E
Cost (relative to single port globe)	0.7	0.9	0.6	0.9	3.0	1.0	1.2	1.1	1.0	0.5	0.7	0.9	0.6	1.0	2.0	1.5
Cryogenic service	A	S	A	A	A	A	A	A	A	X	A	S	X	A	X	X
High pressure drop (over 200 PSI)	A	A	X	A	E	G	G	E	A	X	A	A	X	X	E	E
High temperature (over 200 °C)	Y	S	E	G	Y	Y	Y	Y	Y	X	S	S	X	X	S	X
Leakage (ANSI class)	V	IV	I	IV	V	IV	II	IV	IV	IV	IV	IV	V	I	IV	II
Liquids: abrasive service	C	C	X	X	P	G	G	E	G	G, E	F, G	F, G	F, G	X	E	G
Cavitation resistance	L	L	L	L	M	H	H	H	M	X	L	L	X	L	H	M
Dirty service	G	G	F	G	X	F, G	F	G	F, G	E	G	G	G, E	G	F	F
Flashing applications	P	P	P	F	F	G	G	E	G	F	P	P	F	P	G	P
Slurry including fibrous service	G	G	F	F	X	F, G	F, G	G, E	F, G	E	G	G	E	G	P	E
Viscous service	G	G	G	G	F	G	F, G	G, E	F, G	G, E	G	G	G, E	F	F	F
Gas/vapor: abrasive/erosive	C	C	F	F	P	G	G	E	F, G	G, E	F, G	F, G	G	X	E	E
Dirty	G	G	G	G	X	G	F, G	G	F, G	G	G	G	G	G	F	G
Listing of abbreviations																
A = Available				C = All ceramic design is available				F = Fair				G = Good				
E = Excellent				H = High				L = Low				M = Medium				
P = Poor				S = Special design only				Y = Yes				X = Not available				

Table 3.2
Valve selection orientation table

3.4 Control valve actuators

An actuator is the part of a valve assembly that responds to the output signal of the process controller, causing a mechanical motion to occur which, in turn, results in modification of fluid motion through the valve.

An actuator has to be able to perform two basic functions:

1. To respond to an external signal and cause a valve to move accordingly and with correct selection, other functions can be integrated into this assembly, such as desired fail-safe actions.
2. To provide support (if required) for accessories such as positioners, limit switches, solenoid valves and local controllers.

There are five basic forms of valve actuator, as listed below, and a description of each follows in Section 3.4.1:

1. Digital
2. Electric
3. Hydraulic
4. Solenoid
5. Pneumatic.

The first four have a totally different method of operation and application use as compared to the last one, the pneumatic actuator.

3.4.1 Digital, electric, hydraulic and solenoid actuators

This section describes the common factors of these valves:

1. Digital
2. Electromechanical
   - Stepping motors in smaller size valves
   - Reversible motors and gearboxes for larger size valves
3. Electrohydraulic (the pump being driven by stepping or servo motors)
4. Solenoid operation.

Energy sources

Electrical or electrohydraulic.

Speed reduction techniques

Worm gear, spur gear or gearless.

Torque ranges

- 0.5–30 ft lb (0.6–40 Nm) for type 2a above
- 1–75 000 ft lb (1.3–100 000 Nm) for type 2b.

Speeds of rotation

From 5 to 300 s for a complete opening or closing cycle.

Linear thrust ranges

- Maximum of 500 lb (225 kg) output force from type 2a actuators
- 100–10 000 lb (45–4500 kg) output force from type 2a actuators
- 100 000 lb (45 000 kg) output force from type 3 actuators.

Speeds of full stroke

- Small solenoids can close within 8–12 ms
- Throttling solenoids can stroke in about 1 s
- Electromechanical motor-driven valves stroke in 5–300 s
- Electrohydraulic actuators usually move at 0.25 in./s (6 mm/s).

In essence, actuators have to be able to exert some form of higher torque to overcome the resistance of some types of valve opening, usually described as a high breakaway force. Operation in the opposite direction, that is closing a valve, also sometimes requires extra torque to ensure firm seating is obtained; however, some form of spring-retentive clutch is also needed in case a foreign object is trapped in the valve body.

Limit switches, of many types, ranging from reed, IR, to microswitches can also be mounted within the mechanical assembly of these actuators to signal, and thereby prevent, overrun and excessive movements occurring.

3.4.2   Pneumatic actuators

Pneumatic actuators respond to an air signal by moving the valve trim into a corresponding throttling position. There are two basic types, linear and rotary, the specifications of both being listed in Section 3.4.2.1.

Types and applications of pneumatic actuators

(A) Linear

    a1. Spring diaphragm
    a2. Piston

(B) Rotary

    b1. Cylinder with scotch yoke
    b2. Cylinder with rack and pinion
    b3. Dual cylinder
    b4. Spline or helix
    b5. Vane
    b6. Pneumohydraulic
    b7. Air motor
    b8. Electropneumatic.

The above actuators are applicable to the following valve size:

Type a1: 0.5–8 in. (12.5–200 mm)
Type a2: 0.5–16 in. (12.5–400 mm)
Type B: 2–30 in. (50–750 mm).

Maximum actuator pressure rating

Type a1: 60 PSI (414 kPa); some higher
Type a2: 150 PSI (1035 kPa)
Type B: 250 PSI (1725 kPa).

Actuator areas

Type a1: 25–500 sq.in. (0.016–0.323 sq.m)
Type a2 and B: 10–600 sq. in. (0.006–0.38 sq.m)
Bore diameters from 2 to 44 in. (50 mm to 1.1 m)
Strokes up to 24 in. (0.61 m).

Linear thrust (stem force ranges)

Type a1: 200 to 45 000 lbf (100–20 400 kgf)
Type a2: 200 to 32 000 lbf (100–14 500 kgf)
Specials up to 186 000 lbf (84 000 kgf)

Speeds of full stroke

Type a1: 15 s
Type a2: 0.33–6.0 s (8–150 mm/s).

3.4.3   The steady-state equation

In pneumatic spring and diaphragm actuators, valve stem positioning is achieved by a balance of forces on the stem. Referring to Figure 3.22 (*Forces on a spring-and-diaphragm valve*) the following equation can be derived from a summation of the forces involved, adopting a positive direction downward (closing), and flow is left-to-right, Figure 3.22a:

$$PA - KX - P_V A_V = 0$$

With a reverse flow, right-to-left, Figure 3.22a:

$$PA - KX + P_V A_V = 0$$

The inverse of this, where the stem is moving in a negative direction upwards (opening), and flow is left-to-right, Figure 3.22b:

$$-PA + KX - P_V A_V = 0$$

With a reverse flow, right-to-left, Figure 3.22b:

$$-PA + KX + P_V A_V = 0$$

Where
   A is the effective diaphragm area
   A_V is the effective inner valve area
   K is the spring rate
   P is the diaphragm pressure
   P_V is the valve pressure drop ΔP
   X is the stem travel.

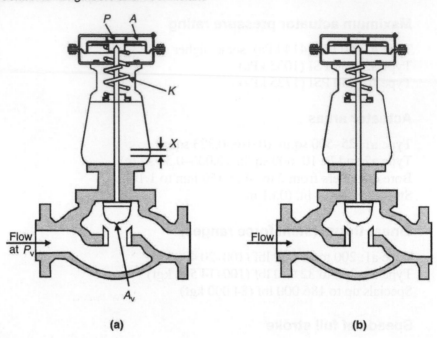

(a) **(b)**

Figure 3.22
Forces on a spring-and-diaphragm forward and reverse acting valve

These equations are simplified because they do not consider friction occurring in the valve stem packing, in the actuator guide and in the valve plug guide(s) or inertia.

Figure 3.23 serves to illustrate the relationship between pressure on a diaphragm and the amount of travel of the valve stem, showing yet another area that generates non-linearity and distortion.

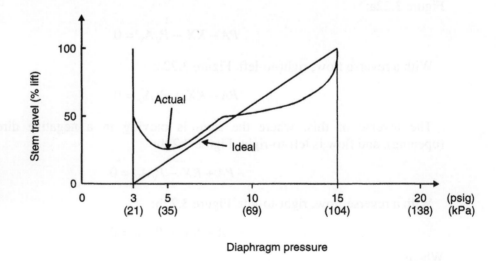

Diaphragm pressure

Figure 3.23
Ideal and actual relationship between diaphragm pressure and valve stem pressure

Table 3.3 indicates the advantages/disadvantages and application for the four most common types of actuator.

Type of Actuator	Advantage	Disadvantage	Application
Linear spring/diaphragm	Low cost	Slow speed	Linear valves
	Mechanical fail-safe	Lack of stiffness	0.5–8 in. Body size
	Moderate thrust	Instability	
	Small package		
	Simple design		
	Excellent control (with or without control devices)		
Linear piston	Low cost	No mechanical fail-safe spring	Linear valves
	Moderate thrust	Slow speed	0.5–16 in. Body size
	Small package	Lack of stiffness	
	Simple design	Instability	
	Excellent control with a control device		
	Long stroke		
Rotary spring/diaphragm	Low cost	Low thrust in spring cycle	Rotary valves 1–6 in. body style
	Mechanical fail-safe	Slow speed	
	Small package	Instability	
	Simple design		
	Easily reversible		
	Excellent control with a control device		
Rotary pistons	Low cost	Slow speed	Rotary valves 1–24 in. body style
	Moderate thrust	Large spring compression	
	Small–large package		
	Mechanical fail-safe		
	Good control with a control device		

Table 3.3
Features of pneumatic actuators

3.5　Control valve positioners

Probably the most significant accessory that can be used for valve control is the positioner, sometimes referred to as 'smart valve electronics' many of which are microprocessor controlled.

A positioner is a high gain proportional controller which measures the stem position, to within 0.1 mm, compares this position to a setpoint, which should be considered as the output of the main process controller, and performs correction on any resultant error signal. The open loop gain of these positioners ranges from 10 to 200 giving a proportional band between 10 and 0.5% and their periods of oscillation ranges from 0.3 to 10 s, a frequency response of $3 - 0.1$ Hz. In other words it is a very sensitive tuned proportional only controller.

The STARPACK system manufactured by Valtek shows, what could be considered a full-house positioner (Figure 3.24).

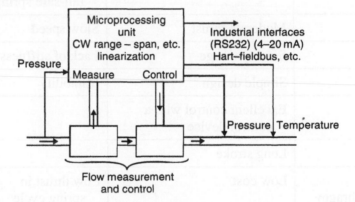

Figure 3.24
Smart valve packages can be provided with local display and sensors for temperature, flow, pressures, pressure differentials and stem position (Courtesy of Valtek)

Not only will it control and measure the flow through the valve, but also measure up and downstream pressures and as such the pressure differential, stem position and temperatures. It has the advantage of being able to store valve 'profiles' to enable software correction or modifications to flow characteristics.

3.5.1　When NOT to use positioners

Remembering that a positioner becomes an intrinsic part of the full control loop very much like the secondary controller in a cascaded system, care must be exercised in their uses. A rule of thumb is that the time constant of the slave should be 10 times shorter (open loop gain 10 times higher), and the period of oscillation of the slave 3 times shorter (frequency response 3 times higher) than that of the primary or master controller.

3.6　Valve sizing

The methods that can be used for the calculations of valve size are many and varied and sometimes very complicated and as such are beyond the scope of this publication. As a rule though, the minimum and maximum C_V requirements for the valve should be determined, and taken into account.

Requirements like '*Process start-up*'; '*Any abnormal process functions required*' and, very importantly '*Reactions required to any Emergency conditions occurring*' must first be taken into account and the valve should be selected to operate adequately over the range of $0.8C_V$ min. to $1.2C_V$ max. If this results in a rangeability which exceeds the capabilities of one valve, then two or more valves should be used.

Control valves should *not* be used outside their rangeability specification. Also, care should be exercised that in summing up all the pressure drops that can occur in a constant pumping speed application, that the result be not applied to the valve for correction, as this always results in *Over sizing* of the valve and as such having it operate for most of its time in a nearly closed position.

4

Fundamentals of control systems

4.1 Objectives

This chapter reviews the basic principles of process control. As a result of studying this chapter, and after having completed the relevant exercises, the student should be able to:

- Clearly explain the concepts of:

  - On–off control
  - Modulating control
  - Open loop control
  - Ratio control.

- List the 10 most common acronyms and basic terminology used in the process control (e.g. PV, MV, OP).
- Describe the differences between a reverse and a direct acting controller.
- Indicate what deadtime is and how it impacts on a process.

4.2 On–off control

The oldest strategy for control is to use a switch giving simple on–off control, as illustrated in Figure 4.1. This is a discontinuous form of control action, and is also referred to as two-position control. The technique is crude, but can be a cheap and effective method of control if a fairly large fluctuation of the process variable (PV) is acceptable.

A perfect on–off controller is 'on' when the measurement is below the setpoint (SP) and the manipulated variable (MV) is at its maximum value. Above the SP, the controller is 'off' and the MV is a minimum.

On–off control is widely used in both industrial and domestic applications. Most people are familiar with the technique as it is commonly used in home heating systems and domestic water heaters. Consider the control action on a domestic gas-fired boiler for example. When the temperature is below the setpoint, the fuel is 'on'; when the temperature rises above the setpoint, the fuel is 'off', as illustrated in Figure 4.2.

There is usually a dead zone due to mechanical delays in the process. This is often deliberately introduced to reduce the frequency of operation and wear on the components. The end result of this mode of control is that the temperature will oscillate about the required value.

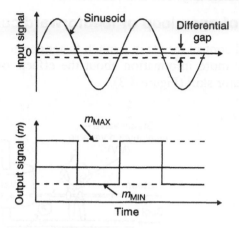

Figure 4.1
Response of a two positional controller to a sinusoidal input

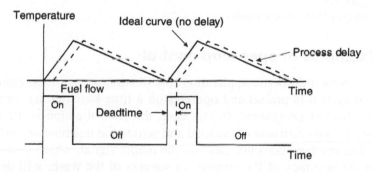

Figure 4.2
Graphical example of on–off control

4.3 Modulating control

If the output of a controller can move through a range of values, this is modulating control.

Modulation control takes place within a defined operating range only, that is, it must have upper and lower limits. Modulating control is a smoother form of control than step control. It can be used in both open loop and closed loop control systems.

4.4 Open loop control

In open loop control, the control action (controller output signal OP) is not a function of the process variable (PV). The open loop control does not self-correct when the PV drifts, and this may result in large deviations from the optimum value of the PV.

4.4.1 Use of open loop control

This control is often based on measured disturbances to the inputs to the system. The most common type of open loop control is feedforward control. In this technique the control action is based on the state of a disturbance input without reference to the actual system condition. i.e. the system output has no effect on the control action, and the input variables are manipulated to compensate for the impact of the process disturbances.

4.4.2   Function of open loop or feedforward control

Feedforward control results in a much faster correction than feedback control but requires considerably more information about the effects of the disturbance on the system, and greater operator skill (Figure 4.3).

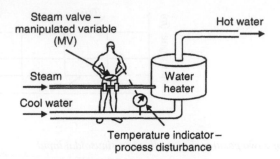

Figure 4.3
Concept of feedforward control

4.4.3   Examples of open loop control

A common domestic application that illustrates open loop control is a washing machine. The system is pre-set and operates on a time basis, going through cycles of wash, rinse and spin as programed. In this case, the control action is the manual operator assessing the size and dirtiness of the load and setting the machine accordingly.

The machine does not measure the output signal, which is the cleanliness of the clothes, so the accuracy of the process, or success of the wash, will depend on the calibration of the system.

An open loop control system is poorly equipped to handle disturbances which will reduce or destroy its ability to complete the desired task. Any control system operating on a time base is an open loop. Another example of this is traffic signals. It is difficult to implement open loop control in a pure form in most process control applications, due to the difficulty in accurately measuring disturbances and in foreseeing all possible disturbances to which the process may be subjected.

As the models used and input measurements are not perfectly accurate, pure open loop control will accumulate errors and eventually the control will be inadequate.

4.4.4   Introduction to ratio control

Ratio control, as its name implies, is a form of feedforward control that has the objective of maintaining the ratio of two variables at a specific value. For example, if it is required to control the ratio of two process variables X_{PV} and Y_{PV} the variable PV_R is controlled rather than the individual PVs (X_{PV} and Y_{PV}).

Thus:

$$PV_R = \frac{X_{PV}}{Y_{PV}}$$

A typical example of this is maintaining the fuel to air ratio into a furnace constant, regardless of maintaining or changing the furnace temperature. This is sometimes known as cross limiting control (Figure 4.4).

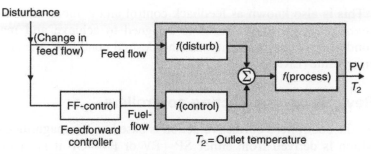

Figure 4.4
Feedforward block diagram

Objective : The objective is to keep the PV constant despite
disturbances. To achieve this, the blocks FF-control
and f(control) must change the PV by the same
magnitude and timing but in opposite direction to
that which the disturbance would have done without
control. Then the feedforward control principle of
compensating the disturbance is fulfilled.

4.5   Closed loop control

In closed loop control, the objective of control, the PV, is used to determine the
control action. The concept of this is shown in Figure 4.5 and the principle is shown in
Figure 4.6.

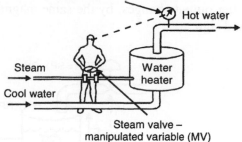

Figure 4.5
Manual feedback control

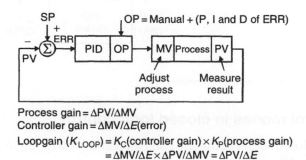

Process gain = ΔPV/ΔMV
Controller gain = ΔMV/ΔE(error)
Loopgain (K_{LOOP}) = K_C(controller gain) $\times$ K_P(process gain)
$\qquad\qquad$ = ΔMV/$\Delta E \times \Delta$PV/ΔMV = ΔPV/ΔE

Figure 4.6
Closed loop block diagram

This is also known as feedback control and is more commonly used than feedforward control. Closed loop control is designed to achieve and maintain the desired process condition by comparing it with the desired condition, the setpoint value (SP), to get an error value (ERR).

4.5.1 Reverse or direct acting controllers

As the controller's corrective action is based on the magnitude-in-time of the error (ERR), which is derived from either SP – PV or PV – SP it is of no concern to the P, I or D functions of the controller which algorithm is used, as the algorithms only change the sign of the error term.

However; if we refer to Figure 4.7 (water level control), which illustrates a controller, performing the same function, but in different ways:

- In case one, we manipulate the *outlet* flow through *V2* to control the tank level; this is *direct* action. Where as the PV increases (tank filling) the OP increases (opening the outlet valve more) to drain the tank faster.

$$\text{Direct acting} = \text{PV} \Uparrow \rightarrow \text{OP} \Uparrow \quad \text{then} \quad \text{ERR} = \text{PV} - \text{SP}$$

- In case two, we control the *inlet* flow through *V1* to control the tank level, this is *reverse* action. Where as the PV increases (tank filling) the OP decreases (closing the inlet valve more) to reduce the filling rate.

$$\text{Reverse acting} = \text{PV} \Uparrow \rightarrow \text{OP} \Downarrow \quad \text{then} \quad \text{ERR} = \text{SP} - \text{PV}$$

The controller output changes, by the same magnitude and sign, based on the resultant error value and sign.

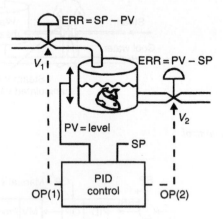

Figure 4.7
Direct and reverse acting controllers

4.5.2 Control modes in closed loop control

Most closed loop controllers can be controlled with three control modes, either combined or separately.

These modes, proportional (P), integral (I) and derivative (D) are discussed in-depth in the next chapter.

4.5.3 Illustration of the concepts of open and closed loop control

The diagrams in Figures 4.4 and 4.6 illustrate the concepts of open loop and closed loop controls in a water heating system.

- In the open loop, feedforward example, the steam flow rate is varied according to the temperature of the cool water entering the system. The operator must have the skills to determine what change in the valve position will be sufficient to bring the cool water entering the system to the desired temperature when it leaves the system.
- In the closed loop, feedback example, the steam flow rate is varied according to the temperature of the heated water leaving the system. The operator must determine the difference between that measurement and the desired temperature and change the valve position until this error is eliminated.
- The above example is for manual control but the concept is identical to that used in automatic control, which should allow greater accuracy of control.

4.5.4 Combination of feedback and feedforward control

The advantages of feedback control are its relative simplicity and its potentially successful operation in the event of unknown disturbances. Feedforward control has the advantage of faster response to a disturbance in the input which may result in significant cost savings in a large-scale operation.

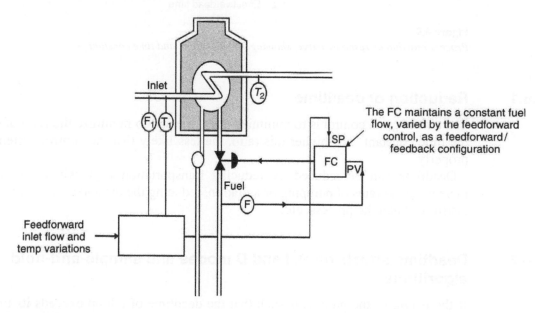

The FC maintains a constant fuel flow, varied by the feedforward control, as a feedforward/feedback configuration

Figure 4.8
Block diagram of feedforward and feedback combination

In general, the best industrial process control can be achieved through the combination of both open and closed loop controls. If an imperfect feedforward model corrects for 90% of the upset as it occurs and the remaining 10% is corrected by the bias generated by the feedback loop, then the feedforward component is not pushed beyond its abilities, the load on the feedback loop is reduced, and much tighter control can be achieved.

4.6 Deadtime processes

In processes involving the movement of mass, deadtime is a significant factor in the process dynamics. It is a delay in the response of a process after some variable is changed, during which no information is known about the new state of the process. It may also be known as the transportation lag or time delay.

Deadtime is the worst enemy of good control and every effort should be made to minimize it. All process response curves are shifted to the right by the presence of deadtime in a process (Figure 4.9). Once the deadtime has passed, the process starts responding with its characteristic speed, called the process sensitivity.

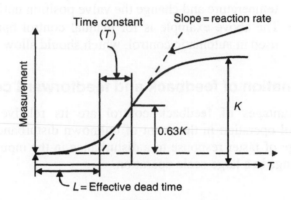

Figure 4.9
Process reaction or response curve, showing both deadtime and time constant

4.6.1 Reduction of deadtime

The aim of good control is to minimize deadtime and to minimize the ratio of deadtime to the time constant. The higher this ratio, the less likely that the control system will work properly.

Deadtime can be reduced by reducing transportation lags, which can be done by increasing the rates of pumping or agitation, reducing the distance between the measuring instrument and the process, etc.

4.6.2 Deadtime effects on P, I and D modes and sample-and-hold algorithms

If the nature of the process is such that the deadtime of a loop exceeds its time constant then the traditional PID (proportional-integral-derivative) control is unlikely to work, and a sample and hold control is used. This form of control is based on enabling the controller so that it can make periodic adjustments, then effectively switching the output to a hold state and waiting for the process deadtime to elapse before re-enabling the controller output. The algorithms used are identical to the normal process control ones except that they are only enabled for short periods of time. Figure 4.10 illustrates this action.

The only problem is that the controller has far less time to make adjustments, and therefore it needs to do them faster. This means that integral setting must be increased in proportion to the reduction in time when the loop is in automatic.

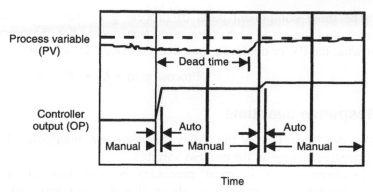

Figure 4.10
Sample and hold algorithms are used when the process is dominated by large deadtimes

4.7 Process responses

The dynamic response of a process can usually be characterized by three parameters: process gain, deadtime and process lag (time constant) (Figure 4.11).

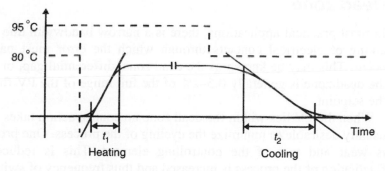

Figure 4.11
Example of a process response related to a step change of the input value

Sections 4.7.1, 4.7.2 and 4.7.3 define the three constitutional parts of the process response curve as illustrated in Figure 4.10.

4.7.1 Response process gain

The process gain is the ratio of the change in the output (once it has settled to a new steady state) to the change in the input. This is the ratio of the change in the process variable to the change in the manipulated variable. It is also referred to as the process sensitivity as it describes the degree to which a process responds to an input.

A slow process is one with low gain, where it takes a long time to cause a small change in the MV. An example of this is home heating, where it takes a long time for the heat to accumulate to cause a small increase in the room temperature. A high gain controller should be used for such a process.

A fast process has a high gain, i.e. the MV increases rapidly. This occurs in systems such as a flow process or a pH process near neutrality where only a droplet of reagent will cause a large change in pH. For such a process, a low gain controller is needed.

The three component parts of process gain from the controllers perspective is the product of the gains of the measuring transducer (K_S), the process itself (K_C) and the gain of what the PV or controller output drives (K_V). This becomes:

$$\text{Process gain} = K_s \times K_c \times K_v$$

4.7.2   Response deadtime

The deadtime (L) is the delay between the manipulated variable changing and a noticeable change in the process variable.

Deadtime exists in most processes because few, if any, real world events are instantaneous. A simple example of this is a hot water system. When the hot tap is switched on there will be a certain time delay as hot water from the heater moves along the pipes to the tap. This is the deadtime.

4.7.3   Response process lag

The process lag (T) is caused by the system's inertia and affects the rate at which the process variable responds to a change in the manipulated variable. It is equivalent to the time constant.

4.8   Dead zone

In most practical applications, there is a narrow bandwidth due to mechanical friction or arcing of electrical contacts through which the error must pass before switching will occur. This may be known as the dead zone, differential gap, or neutral zone. The size of the dead zone is generally 0.5–2% of the full range of the PV fluctuation, and it straddles the setpoint.

When the PV lies within the dead zone no control action takes place, thus its presence is usually desirable to minimize the cycling of the process. One problem with on–off control is wear and tear of the controlling element. This is reduced as the bandwidth of fluctuation of the process is increased and thus frequency of switching decreased.

Exercise 1 (p.231)

Single Flow Loop – Flow Control Loop Basic Example

This will give practical experience in the concept of closed loop control. It would be appropriate to do this exercise now, in order to become familiar with the concepts of closed loop control as well as the operation of the simulation software.

5

Stability and control modes of closed loops

5.1 Objectives

As a result of studying this chapter, and after having completed the relevant exercises, the student should be able to:

- Indicate what stability is, and mathematically what causes instability
- Describe the function and use of proportional, integral and derivative control and various combinations of these terms
- Indicate what problems in closed loop control are caused by and how to correct them.

5.2 The industrial process in practice

We have seen the basic principles of closed loop control in the previous chapter. A control action is calculated, based on the deviation of the PV from the desired value of control as defined by the SP (ERR = PV – SP).

We have to consider the industrial process as it works in the real world. As an example of this, which we will now review, is a feed heater which is used to heat up material before it is fed into a distillation column (see Figure 5.1).

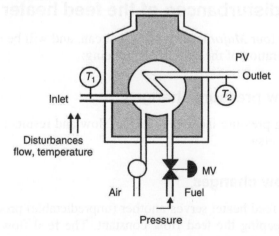

Figure 5.1
Temperature control of a feed heater

The objective of the system is temperature control of the outlet temperature (T_2) that should be kept constant. The manipulated variable is the fuel valve position.

It should be noted, that for economic and environmental reasons, cross limiting control of the combustion is normally required to minimize the output of carbon monoxide. In this example for simplicity, we will neglect cross limiting control totally and manipulate the valve position directly.

This example of feed heater control will serve as an example for us to look into the practical implications of stability, different control modes, control strategies and practical exercises. For this reason we will first have a closer look into the basic dynamic behavior and the most common disturbances of the process which affect this control system.

5.3   Dynamic behavior of the feed heater

There are two major types of systems lag, control and disturbance, that effect the dynamic behavior of this heater system.

5.3.1   Control lag

A lag between positioning of the fuel valve and the outlet temperature exists. The main reason for this lag can be seen by virtue of the fact that not all feed material in the heater will be heated up at the same time after a change of the fuel valve position. Some part of the feed material in the heater at the time of fuel valve change will leave the heater shortly after and some other part later. A minor deadtime is also a part of the control reaction.

5.3.2   Disturbance lags

The impact of disturbances on the outlet temperature also has a lag action. Every disturbance has its own lag time constant. Most disturbances have a minor deadtime as well.

Note: There is no measurable difference between two high order lags one with a minor deadtime and the other without.

5.4   Major disturbances of the feed heater

There are four *Major* disturbances that can, and will be considered as being critical to the stable operation of the system, these being:

5.4.1   Fuel flow pressure changes

Increasing pressure increases the fuel flow and results in a higher outlet temperature (T_2) and vice versa.

5.4.2   Feed flow changes

Since the feed heater serves another (unpredictable) process downstream of it, there is no way of keeping the feed flow constant. The feed flow depends totally on the need for material by the following process. An increase in the feed flow (demanded by the downstream process) decreases the outlet temperature and vice versa.

5.4.3 Feed inlet pressure changes

If the feed material is in the form of gas, this becomes an important issue. It is important to know the mass-flow rather than the volumetric flow of the feed material. With increasing pressure we increase the mass flow which results in a decrease of the outlet temperature and vice versa.

5.4.4 Feed inlet temperature changes

The higher the inlet temperature, the less we have to heat. An increase in inlet temperature results in an increase of the outlet temperature and vice versa.

5.5 Stability

We have stability in a closed loop control system if we have no continuous oscillation. We must not confuse the problems and the different effects that disturbances, noise signals and instability have on a system. A noisy and disturbed signal may show up as a varying trend, but it should never be confused with loop instability.

The criteria for stability are these two conditions:

1. The loop gain (K_{LOOP}) for the critical frequency <1
2. Loop phase shift for the critical frequency <180°.

5.5.1 Loop gain for critical frequency

Consider the situation where the total gain of the loop for a signal with that frequency has a total loop phase shift of 180°. A signal with this frequency is decaying in magnitude, if the gain for this signal is below 1. The other two alternatives are:

1. Continuous oscillations which remain steady (loop gain = 1)
2. Continuous oscillations which are increasing, or getting worse (loop gain >1).

5.5.2 Loop phase shift for critical frequency

Consider the situation where the total phase shift for a signal with frequency that has a total loop gain of 1. A signal with this phase shift of 180° will generate oscillations if the loop gain is greater than 1. This situation is illustrated in Figure 5.2.

Note:

- Increasing the gain or phase shift destabilizes a closed loop, but makes it more responsive or sensitive.
- Decreasing the gain or phase shift stabilizes a closed loop at the expense of making it more sluggish.
- The gain of the loop (K_{LOOP}) determines the offset value of the controller and offset varies with setpoint changes.

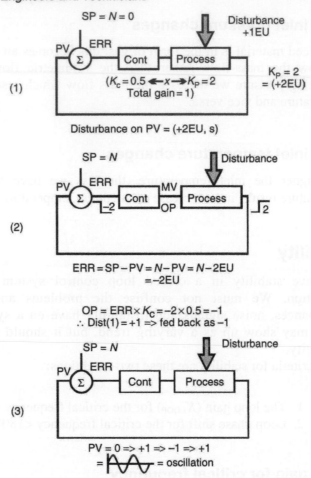

Figure 5.2
Increasing Instability with a 180° phase shift (and gain = >1)

5.6　Proportional control

This is the principal means of control. The automatic controller needs to correct the controllers OP, with an action proportional to ERR. The correction starts from an OP value at the beginning of automatic control action.

5.6.1　Proportional error and manual value

We will call this starting value MANUAL. In the past, this has been referred to as 'manual reset'. In order to have an automatic correction made, that means correcting from the MANUAL starting term, we always need a value of ERR. Without an ERR value there is no correction and we go back to the value of MANUAL. We therefore always need a small 'left over' error to keep the corrective control up. This left over error is called the offset. ERR0 is the error value we would have without any control at all.

K_C is the gain applied to scale the size of the control action based on ERR. LOOP is the total loop gain which is the product of controller gain (K_C) and process gain (K_P). The only tuning constant for proportional control is K_C (controller gain). The larger we make the value of K_C, the more difficult or sensitive (reduced stability) is the control of the system.

With larger values of K_C, the offset value becomes smaller. If the gain is made too large, we may face a stability problem. The following relationships follow from the above:

5.6.2 Proportional relationships

1. $OP = K_C \times ERR + MANUAL$
2. $K_{LOOP} = K_C \times K_P$
3. $Offset = ERR0 / (K_{LOOP} + 1)$

$$ERR = SP - PV$$
$$OP = K_C \times ERR$$
$$PV = ERR \times K_{LOOP}$$
$$ERR = SP - PV$$
$$= SP - ERR \times K_{LOOP}$$
$$\therefore ERR + ERR \times K_{LOOP} = SP$$
$$ERR (1 + K_{LOOP}) = SP$$

At a steady state
$$ERR = SP/(1 + K_{LOOP})$$

The error term (ERR) is defined as 'error = Indicated – Ideal' and is produced as:

$$ERR (t) = SP (t) - PV (t)$$

Although this indicates that the setpoint (SP) can be time-variable, in most process-control problems it is kept constant for long periods of time. For a proportional controller the output is proportional to this error signal, being derived as:

$$OP_C (t) = P + K_C E(t)$$

Where

OP_C = The controller output
P = The controller output bias, or MANUAL starting value
K_C = The controller gain (usually dimensionless)
E = The ERROR value.

This leads the way to evaluating a set of concepts for proportional control.

5.6.3 Evaluation of proportional control concepts

- The controller gain (K_C) can be adjusted to make the controller output (OP_C), changes as sensitive as desired to differences that occur between the SP and PV values.
- The sign of K_C can be chosen (+ or –) to make OP_C either increase or decrease as the deviation or ERR value increases.

In proportional controllers, the MANUAL or starting value of the OUTPUT is adjustable. Since the controller output equals the value of MANUAL when the error value is zero (SP = PV), the value of MANUAL is adjusted so that the controller output and consequently the manipulated variable, MV, are at their nominal steady-state values.

For example, if the controller output drives a valve, MANUAL is adjusted so that the flow through the valve is equal to the nominal steady-state value when ERR = 0.

The gain K_C is then adjusted and for general controllers it is dimensionless that is the terms MANUAL and ERR have the same unit terms of measurement.

The disadvantage of proportional controllers is that they are unable to eliminate the steady-state errors that occur after a setpoint or a sustained load change.

5.6.4 Proportional band

A controllers proportional band is usually defined, in percentage terms, as the ratio of the input value, or PV to a full or 100% change in the controller output value or MV. Its relationship to proportional, or controller gain (K_C) is given by:

$$PB = \frac{1}{K_C} \times 100$$

Proportional:

$$\Delta MV = K_C \times \Delta PV$$

Proportional band %:

$$\Delta PV = \frac{\Delta MV}{K_C}$$

when $\Delta MV = 100\%$.

As shown in Figure 5.3, if the PB, or proportional band, of a controller is set at 100% ($K_C = 1$) then a full change of the PV, or input, from 0 to 100% will result in a change of the MV, or output, from 0 to 100%, resulting in 100% of valve motion or operation.

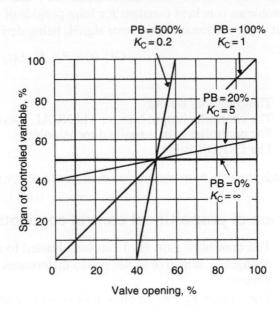

Figure 5.3
Ranges of proportional bands

If the PB is set at 20% ($K_C = 5$) then a change in the PV, or input, from 40 to 60% will result in the same change of the MV, or output, from 0 to 100%. With the same resultant motion of the valve from fully closed to fully open. Likewise, a PB value of 500% ($K_C = 0.2$) will result in the MV, or output, changing from 40 to 60% when the PV, or input, changes from 0 to 100%.

High percentage values of the PB therefore constitute a less sensitive response from the controller while low percentage values result in a more sensitive response.

Exercise 2 (p. 234)

Single Flow Loop – Proportional (P) Control ~ Flow Control

This exercise will introduce the main control action of controllers – proportional control.

5.7   Integral control

Integral action is used to control towards no offset in the output signal. This means that it controls towards *no error* (ERR = 0). Integral control is normally used to assist proportional control. We call the combination of both PI-control.

5.7.1   Integral and proportional with integral formula

Formula for I-control:

$$OP = \left(\frac{K}{T_{INT}} \right) \int_O^T ERR \, dt$$

Formula for PI-control:

$$OP = \left(\frac{K}{T_{INT}} \right) \int_O^T ERR \, dt + (K \times ERR + MANUAL)$$

T_{INT} is the integral time constant.

Since integral control (I-control) integrates the error over time, the control action grows larger the longer the error persists. This integration of the error takes place until no error exists. Every integral action has a phase lag of 90° This phase shift has a destabilizing effect. For this reason, we rarely use I-control without P-control.

5.7.2   Integral action

Let us review a few principles of calculus and trigonometry in relation to integral calculation, especially the integration of a sine wave. Figure 5.4 shows the phase lag of the integral calculation on a sine wave. The same effect exists if integral action is used in a closed loop control system. The integral action adds to the existing phase lag. The maximum of the integrated sine wave is when the sine wave swings back.

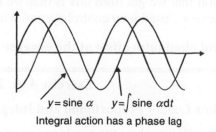

$y = \text{sine } \alpha \qquad y = \int \text{sine } \alpha dt$

Integral action has a phase lag

Figure 5.4
The phase shift of the integration action

If we consider a 'steady-state' value exists for the ERR term, then the integral output will, at the completion of each of its time constants, T_{INT}, increase its output value by ERR $\times K_C$ in the form of a ramp as shown in Figure 5.5(a).

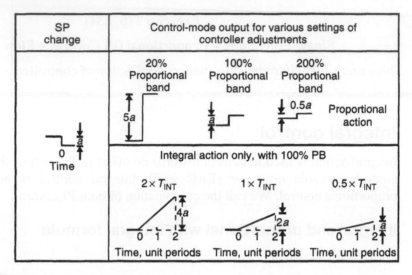

Figure 5.5(a)
Integral relationships and output

5.7.3   Integral action in practice

In practice, as the integral output increases and passes through the process the PV will move towards the SP value and the ERR term will reduce in magnitude. This will reduce the rate-of-change during the integral time interval, resulting in the classic first-order 'curve' response shown in Figure 5.5(b).

If the rate-of-change or the value on T_{INT} is too small, along with the 90° phase lag in the integral action, oscillations may occur, i.e., in effect, applying *over-correction-in-time* to the value of the offset term.

If this happens with a closed loop control system in the industry, we have a stability problem.

Exercise 3 (p. 237)

Single Flow Loop – Integral (I) Control ~ Flow Control

This exercise will introduce the integral control action of controllers.

The conclusion that we get from this is that we have to be careful in the use of integral control if we have a closed loop control system which has a tendency towards instability.

Integral control eliminates offset at the expense of stability

---

Exercise 4 (p. 240)

Single Flow Loop – Proportional and Integral (PI) Control ~ Flow Control

This exercise will introduce the combination of the proportional and integral control action of controllers.

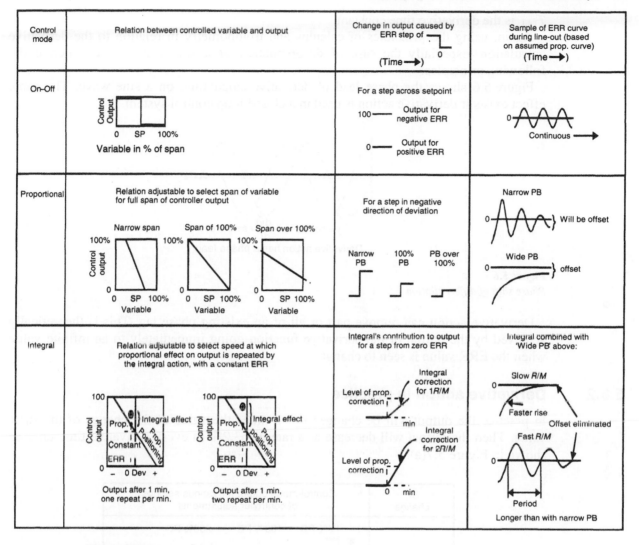

Figure 5.5(b)
Integral action in practice

5.8 Derivative control

The only purpose of derivative control is to add stability to a closed loop control system.

The magnitude of derivative control (D-control) is proportional to the rate of change (or speed) of the PV.

Since the rate of change of noise can be large, using D-control as a means of enhancing the stability of a control loop is done at the expense of amplifying noise. As D-control on its own has no purpose, it is always used in combination with P-control or PI-control. This results in a PD-control or PID-control. PID-control is mostly used if D-control is required.

5.8.1 Derivative formula

Formula for D-control:

$$OP = K \times T_{DER} \left(\frac{dERR}{dt} \right)$$

T_{DER} is the derivative time constant.

Again, using the principles of calculus and trigonometry in relation to the derivative calculation, especially the case of differentiation of a sine wave we can derive the following principles.

Figure 5.6 shows the phase lead of derivative calculation on a sine wave. The same effect exists if derivative action is used in a closed loop control system.

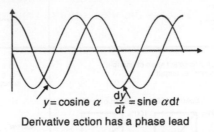

$y = \text{cosine } \alpha \qquad \dfrac{dy}{dt} = \text{sine } \alpha \, dt$

Derivative action has a phase lead

Figure 5.6
Phase shift of differentiation

Derivative action can remove part or all of an existing phase lag. This is theoretically achieved by the output of the derivative function going immediately to an infinite value when the ERR value is seen to change.

5.8.2   Derivative action in practice

In practice the output will be changed to +8 times the value of the change of the ERR value. Then the output will decrease at a rate of 63.2% in every derivative time unit, as shown in Figure 5.7(a).

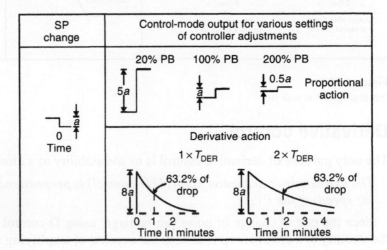

Figure 5.7(a)
Derivative relationships and output

5.8.3   Summary of integral and derivative functional relationships

Integration can be considered as charging a capacitor, from a constant voltage source, via a resistor. The voltage across the capacitor rises from a zero value in an exponential form. This being caused by the difference between the supply and capacitor voltage reducing in time (Figure 5.7(b)).

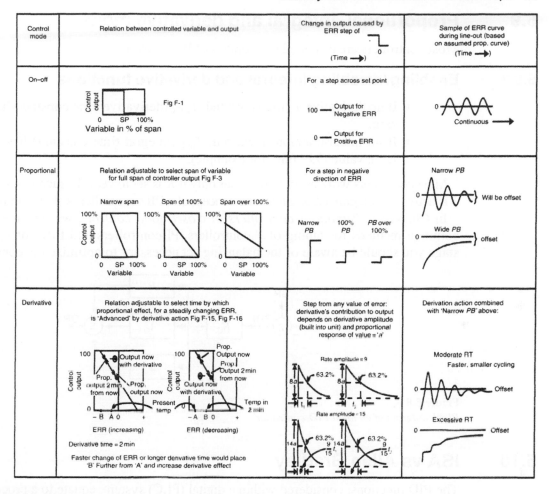

Figure 5.7(b)
Derivative control has no functionality on its own

Derivative action is in essence the inverse of the example for integral action. Taking a fully charged capacitor and discharging it through a resistor results in an exponential decay, as the difference in capacitor voltage reduces from its maximum value to zero.

At first glance, it would appear that the integral and derivative functions, one being the inverse of the other, would effectively cancel out each other. However it has to be remembered that the ERR term is dynamic and constantly changing.

There is a fairly strict ratio between T_{INT} and T_{DER} and the process or loop time T_{PROC}. these relationships being explained in Chapter 8 (under section 'system tuning procedures').

Exercise 5 (p. 242)

General Single Loop with Interactive PID (Real Form) – Introduction to Derivative (D) Control

This exercise will introduce the Derivative Control action of controllers

5.9 Proportional, integral and derivative modes

Most controllers are designed to operate as PID-controllers.

5.9.1 Enabling/disabling integral and derivative functions

- If no derivative action is wanted, T_{DER} (derivative time constant) has to be set to zero.
- If no integral action is wanted, T_{INT} (integral time constant) has to be set to a large value (999 min, for example).

Most controllers work as an I-controller only if K is set to zero. In such cases, a unit gain of 1 is active for integral action only. The concept of a PID-controller is shown in Figure 5.8.

In Chapter 8 *'Tuning of controllers in closed loop control'*, we will review the most common methods for tuning of P-controllers, PI-controllers and PID-controllers. At this stage you should be aware of the balancing act necessary to optimize the control action.

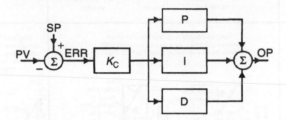

Figure 5.8
Block diagram of an ideal PID-controller

5.10 ISA vs Allen Bradley

The PID functions, considered within a digital (PLC) system, equate to a process where the output of a controller is designed to drive the process variable (PV) toward the setpoint (SP) value. The difference between the PV and SP values is the system error value, upon which the PID functions operate. The greater the error value the greater the output signal.

ISA (Instrument Society of America) has a set of rules that make the P, I and D functions dependent on each other, and for example, the Allen Bradley PLC system operates either on ISA (dependent) or independent gains.

Chapter 8 illustrates the differences.

5.11 P, I and D relationships and related interactions

P-control is the principle method of control and should do most of the work.

I-control is added carefully just to remove the offset left behind by P-control.

D-control is there for stability only. It should be set up so that its stabilizing effect is larger than the destabilizing effect of I-control.

In cases where there is no tendency towards instability, D-control is not used. This includes most flow applications.

Exercise 6 (p. 246)

Practical Introduction into Stability Aspects

Gives practical experience on the topics of closed loop stability.

5.12 Applications of process control modes

5.12.1 Proportional mode (P)

The most basic form of control. This can be used if the resultant offset in the output is constant and acceptable. Varied by the controller gain K_C.

5.12.2 Proportional and integral mode (PI)

Integral control can be added to the proportional control to remove the offset from the output. This can be used if there are no stability problems such as in a tight flow control loop.

5.12.3 Proportional, integral and derivative mode (PID)

This is a full 3-term controller, used where there is instability caused by the integral mode being used. The derivative function amplifies noise and this must be considered when using the full three terms.

5.12.4 Proportional and derivative mode (PD)

This mode is used when there are excessive lag or inertia problems in the process.

5.12.5 Integral mode (I)

This mode is used almost exclusively in the primary controller in a cascaded configuration. This is to prevent the primary controllers output from performing a 'step change' in the event of the controllers setpoint being moved.

5.13 Typical PID controller outputs

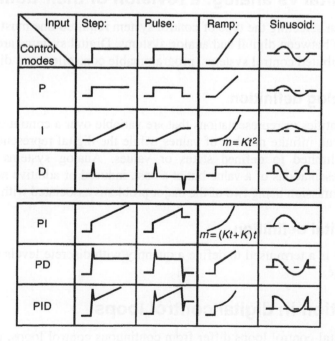

Figure 5.9
Typical controller outputs

6

Digital control principles

6.1   Objectives

As a result of studying this chapter, and after having completed the relevant exercises, the student should be able to:

- Identify and describe the mathematical form of the most important building blocks used in industrial control
- Describe the principles applied in computer-based digital controllers
- Indicate what a real time program is.

In order to best understand the control algorithms used in industrial control, it is appropriate to look at the building blocks first.

6.2   Digital vs analog: a revision of their definitions

When selecting the type of control system required one must examine the alternatives that exist between digital and analog systems. Digital systems are compatible with computers, distributed control systems, programmable controllers and digital controllers.

6.2.1   Analog definition

Quantities or representations that are variable over a continuous range. These variables can take an infinite number of values, while the digital representation of these same variables are limited to defined states or values. Analog systems are more accurate in their representations of a value but at a cost, induced or additive noise and difficulty in accurate transmission being two of the major problems associated with this type of system.

6.2.2   Digital definition

This is a term used to define a quantity with discrete levels rather than over a continuous range.

6.3   Action in digital control loops

Digital control loops differ from continuous control loops, their analog cousins, in that a continuous controller is replaced by a *sampler*. This is some form of a computer performing discrete control algorithms and storing the individual results.

Action is based on comparing the difference between previous sampled value(s) and the current value and generating an output which is used to increment or decrement the final controller output, in conjunction with any other existing digital function (P or P + I or P + I + D, etc.).

6.4   Identifying functions in the frequency domain

As control algorithms are often expressed in terms of $f(s)$ which refers to a function in the *frequency domain*, we will review these expressions.

This paragraph is not intended to go into the theory of the Laplace transforms, but to provide a basic understanding of the expressions needed to understand the composition of most control algorithms. However a quick and simple revision and overview follows.

6.4.1   Laplace conceptual revision

The principle of a transform operation is to change a difficult problem into an easier problem or form that is more convenient to handle. Once the result from a transformation has been obtained an inverse transformation can be made to determine the solution to the original problem. For example, logarithms are a transform operation by which problems of multiplication and division can be transformed into summing and negation operations.

Laplace transforms perform a similar function in the solution of differential equations. The Laplace transform of a *linear ordinary differential equation* results in a *linear algebraic equation*. This is usually much more simpler to solve than the corresponding differential equation. Once the Laplace domain solution has been found, the corresponding *time domain* solution can de determined by using an inverse transformation.

The Laplace function of a time domain function $f(t)$ is denoted by the symbol $F(s)$ and is defined as follows:

$$F(s) = L[f(t)] = \int_0^\infty f(t)et^{-st}\,\mathrm{d}t$$

Where

$L[f(t)]$ is the symbol for the Laplace transformation in the brackets

The variable s is a complex variable $(s = a + jb)$ introduced by the transformation.

All time dependant functions in the time domain become functions of s in the Laplace domain (s domain).

The following example illustrates an integrator as an integral block with its step function input $1/s$ in the frequency domain being represented as an integral calculation.

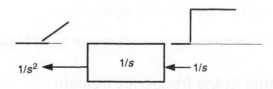

Appendix A illustrates some of the Laplace transform pairs.

6.4.2 Common building blocks

The most commonly used building blocks are:

- Ts: Derivative block with derivative time constant
- $1/Ts$: Integral block with integral time constant
- $1/(1 + Ts)$: First order lag with lag time constant block.

We can work with these blocks using the block diagram transformation theorems also referred to as *block diagram algebra.*

An example of this is the building of a lead algorithm. The lead algorithm is the derivative of a lag algorithm, where the derivative time constant (T_{DER}) has to be significantly larger than the lag time constant (T_{lag}).

$$Lead = Derivative \times lag$$

$$Lead = s\,T_{DER} \times \frac{1}{(1 + s\,T_{lag})}$$

$$Lead = \frac{s\,T_{DER}}{(1 + s \times T_{lag})}$$

Approaching the problem from the other direction, we will analyze existing control algorithms by building block diagrams with blocks using the above terms. Then we will review the way these blocks are implemented in digital computers.

In Figure 6.1 we see the block diagram of a *real* controller used as an ultimate secondary, or field controller, driving the actual variable of the process.

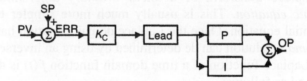

Figure 6.1
Field (real) controller block diagram

The formula in terms of $f(s)$ for the control algorithm of controllers, based on the block diagram in Figure 6.1 can be stated as:

$$OP = K \times \frac{1 + T_{DER}\,s}{1 + \alpha\,T_{DER}\,s} \times \frac{1 + T_{INT}\,s}{T_{INT}\,s}$$

Where

$$K = \text{Controller gain}$$
$$T_{INT} = \text{Integral time constant}$$
$$T_{DER} = \text{Derivative time constant (lead} = \alpha \text{ times lag)}$$
$$Alpha = \alpha\,(\alpha = 8 \text{ for training applications}).$$

Industrial controllers use a value between 8 and 12 for $1/\alpha$.

6.4.3 Algorithms in the frequency domain

Algorithms expressed in the frequency domain do not show any static constants. Therefore, the algorithms have to be calculated independently of any constant. For example, such a constant could be the manual starting position of an OP value. This coincides with the need to have all dynamic control calculations made to be *independent of the absolute value of* OP. The requirement is there because the OP value has to be modified from the destination (the slave controller) of the value if the destination is capable of initialization.

We will review initialization in the Chapter on 'Cascade control'. If no initialization takes place, the OP value is calculated by the controller algorithms (automatic control). Every time we change from the initialization state into automatic control, the OP value has to be accepted as it is. Otherwise there would be a 'bump' in the OP value in changing from the initial manual state into automatic mode which could cause a process upset.

6.5 The need for digital control

There is a requirement to modify the OP value from different independent calculations like initialization and automatic control, and so neither of these calculations must have control over the absolute value of OP. These calculations are allowed to *increment and decrement an existing* OP *value only*. They *do not determine* the absolute value of OP. Therefore the absolute value of OP reflects the destination value only.

6.5.1 Incremental algorithms

The OP value for example can show the true valve position and no calculation is permitted to force an absolute value on OP. Only changes that means movements of the valve positions are permitted.

This approach uses what we call an incremental algorithm *where the control calculations calculate* changes *and* not absolute values.

Once this principle is established, it can be used to calculate PID-control in separate:

- P-calculation
- I-calculation and
- D-calculation.

each incrementing (or decrementing) the OP value *without knowing the other control mode calculations*. Every calculation is merely incrementing (or decrementing) the OP and does not care about the absolute value of the OP.

The principle of incremental OP calculation for automatic control based on the block diagram in Figure 6.2: the *ideal* controller.

$$OP_n = OP_{n-1} + \Delta OP_P + \Delta OP_I + \Delta OP_D$$

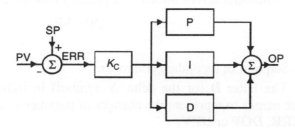

Figure 6.2
Ideal PID controller block diagram

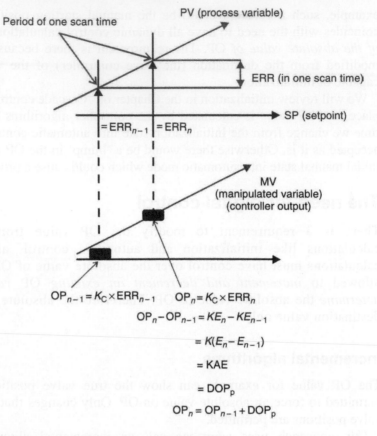

Figure 6.3
Graphical example of DOP_P

The principle of incremental OP calculation for automatic control based on block diagram in Figure 6.1: the *real* controller.

$$OP_n = OP_{n-1} + \Delta OP_P + \Delta OP_I$$

Where
   OP_n = Output value after current scan
   OP_{n-1} = Output value after the last scan time
   ΔOP_P = Change to output value required by the proportional action
   ΔOP_I = Change to output value required by the integral action
   ΔOP_D = Change to output value required by the derivative action.

If in *cascade* control (see Chapter 'Cascade control') and initialization, the SP of a secondary controller drives the OP of a primary controller (Figure 6.3).

$$OP = SP_S$$

Where
   SP_S = Setpoint of secondary controller.

Note: The letter **D** (or the delta Δ symbol) in italics; has been used as prefix for parameter names to represent the changes of parameters from one calculation to the next, as in **DERR**, **DOP** or **DPV**.

The time from one calculation to the next is called the *scan time*.

For full value representation of the parameters, no prefix has been used, as in ERR, OP or PV.

6.6   Scanned calculations

A digital computer cannot perform a number of related calculations simultaneously. A series of repeated calculations is thus made.

- If the repetition interval between calculations is constant, we call it a *fixed scan time*.
- A fixed scan time is used in all controllers designed for *continuous (modulating) control*.
- If the scan time is not constant as with some programmable logic controllers (PLCs), the scan time has to be calculated for each scan of the computer system.
- This is especially important, since all time constants used for the actual scanned (repetitive) calculation have to be used in units of scan.

Therefore to summarize for scanned (repetitive) calculations:

- All time constants are in *units of scan.*
- All time constants must be far greater than the scan time to ensure that the digital calculation is the equivalent, or a good approximation to that of an analog calculation.

6.7   Proportional control

Let us compare the general formula shown before with the formula used for incremental P-control:

$$OP = K \times ERR + MANUAL$$

After differentiation:

$$d\frac{OP}{dt} = K \times \frac{dERR}{dt}$$

Note that we have lost our constant *MANUAL*. This makes this algorithm a *dynamic calculation only*. If the process reaction is insignificant between scan times, we can simplify the calculation into a difference calculation with the interval of scan time:

$$\Delta OP = K \times \Delta ERR$$

ΔERR is the change of error from the last scan to the present scan. ERR in a difference equation is the equivalent of ERR dt in a differential equation.

6.8   Integral control

Let us compare the general formula shown before with the formula used for incremental I-control:

$$OP = \left(\frac{K}{T_{INT}}\right) \int_0^T ERR \, dt$$

After differentiation:

$$\frac{dOP}{dt} = \left(\frac{K}{T_{INT}}\right) \times ERR$$

If the process reaction is insignificant between scan times, we can simplify the calculation into a difference calculation with the interval of scan time:

$$\Delta OP = \frac{K}{T_{INT}} \times ERR$$

Where:

$$T_{INT}[scan\ units] = \left(\frac{T_{INT}[min] \times 60}{scan[s]} \right)$$

Note: T_{INT} has to be in units of scan (or number of scans), not in minutes or seconds. For example, if the interval of repeated calculation (scan time) is 0.5 s and T_{INT} is 1.5 min or 90 s, then T_{INT} in units of scan is 180. Put another way, T_{INT} is 180 units each of 0.5 s duration.

6.9   Derivative control

Let us compare the general formula shown before with the formula used for incremental D-control:

$$OP = K \times T_{DER} \left(\frac{dERR}{dt} \right)$$

After differentiation:

$$dOP/dt = K \times T_{DER} \left(\frac{d^2ERR}{dt^2} \right)$$

If the process reaction is insignificant between scan times, we can simplify the calculation into a difference calculation with the interval of scan time:

$$DOP = K \times T_{DER} \times \Delta(\Delta ERR)$$

Where:

$$T_{DER}[scan\ units] = [T_{DER}[min] \times 60]/SCAN[s]]$$

Note: T_{DER} has to be in units of scan (or number of scans). $\Delta(\Delta_{ERR})$ is the change of the change of error from the last scan to the present scan. $\Delta(\Delta_{ERR})$ in a difference equation is the equivalent of d^2ERR/dt^2 in a differential equation.

6.10   Lead function as derivative control

The real algorithm used for the field controller does not use the idealistic and mathematically simplest approach. Instead of a mathematically defined derivative action, the field controller uses a lead algorithm for derivative control. The formula in terms of $f(s)$ for the control algorithm of a field controller using a lead algorithm is shown in Figure 6.4. The block diagram is shown in Figure 6.5.

$$OP = K \times \frac{T2s + 1}{\alpha T2s + 1} \times \frac{1 + T1s}{T1s}$$

$$= Gain \times Lead \times PI - Control$$

Figure 6.4
Formula for a FIELD controller in terms of F(s) *using a lead algorithm*

The *lead* part, acting for *derivative* control is explained in detail in Figure 6.5 below:

$$\frac{Ts+1}{\alpha Ts+1} = \left(\frac{1}{\alpha Ts+1}\right) \times (Ts+1)$$

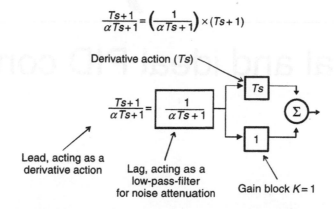

Figure 6.5
Block diagram of lead as derivative

If we consider: $\alpha Ts = 1/8Ts$ then this means derivative is 8 times more powerful than the low-pass-filter. This approach keeps the adverse effect noise has on the derivative term to an acceptable minimum.

6.11 Example of incremental form (Siemens S5-100 V) (Figure 6.6)

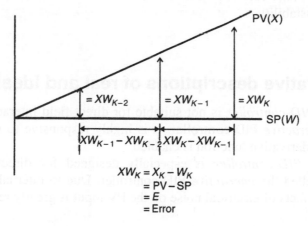

Figure 6.6
Example of an incremental control

Change of output for proportional action $= \Delta OP_P = K(XW_K - XW_{K-1})R$

Change of output for integral action $= \Delta OP_I = KT_{INT}WK$

Change of output for derivative action $= \Delta OP_D = KT_{DER}((XW_K - XW_{K-1})$
$$- (XW_{K-1} - XW_{K-2}))$$
$$= KT_{DER}(XW_K - 2XW_{K-1} + XW_{K-2})$$

$$\therefore d\psi_K = K[(XW_K - XW_{K-1})R + T_{INT}XW_K + T_{DER}(XW_K - 2XW_{K-1} + XW_{K-2})]$$

7

Real and ideal PID controllers

7.1 Objectives

As a result of studying this chapter, and after having completed the relevant exercises, the student should be able to:

- Select the correct PID-control algorithm for field interaction and for computer-optimized calculations
- Clearly distinguish between process noise and control loop instability, which are often similar in appearance
- List the correct sequence of steps to handle the different problems of noise and instability.

7.2 Comparative descriptions of real and ideal controllers

The *ideal PID-controller* is not suitable for direct field interaction, therefore it is called the *non-interactive* PID-controller. It is highly responsive to electrical noise on the PV input if the derivative function is enabled.

The *real PID-controller* is especially designed for direct field interaction and is therefore called the *interactive* PID-controller. Due to internal filtering in the derivative block the effects of electrical noise on the PV input is greatly reduced.

7.3 Description of the ideal or the non-interactive PID controller

The non-interactive form of controller is the classical teaching model of PID algorithms. It gives a student a clear understanding of P, I and D control, since:

P-control, I-control and D-control can be seen independently of each other. Then, PID is effectively a combination of independent P, I and D-control actions. This can be seen in Figure 7.1.

Since P, I and D algorithms are calculated independently in an ideal PID-controller, this form of controller is recommended if an *ideal process variable exists*.

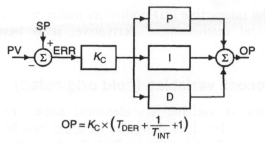

$$OP = K_C \times \left(T_{DER} + \frac{1}{T_{INT}} + 1 \right)$$

OP = Gain × (D-control + I-control + P-control)

Figure 7.1
Ideal PID-controller

7.3.1   Ideal process variables

An ideal process variable is a noise-free, refined and optimized variable. They are a result of computer optimization, process modeling, statistical filtering and value prediction algorithms. These types of ideal process variables do not come from field sensors. In these cases, it is of great benefit that the actual formula of the *Ideal PID algorithm* is simple, as shown in Figure 7.2.

$$OP = K_C \times \left(T_{DER}\, s + \frac{1}{T_{INT}\, s} + 1 \right)$$

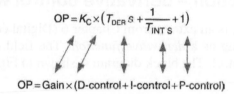

OP = Gain × (D-control + I-control + P-control)

Figure 7.2
Ideal PID algorithm

7.4   Description of the real (interactive) PID controller

The interactive form is the PID algorithm used for direct field control. That is either both of its input (PV) and output (MV) are directly connected to field or process equipment. It is designed to cope with any electrical noise induced into its circuits by equipment in the plant or factory.

Full understanding of the interactive PID algorithm is rather difficult, since P-control, I-control and D-control cannot be seen independently from each other. Therefore, interactive PID is not just a sum of independent P, I and D control. This can be seen in Figure 7.3.

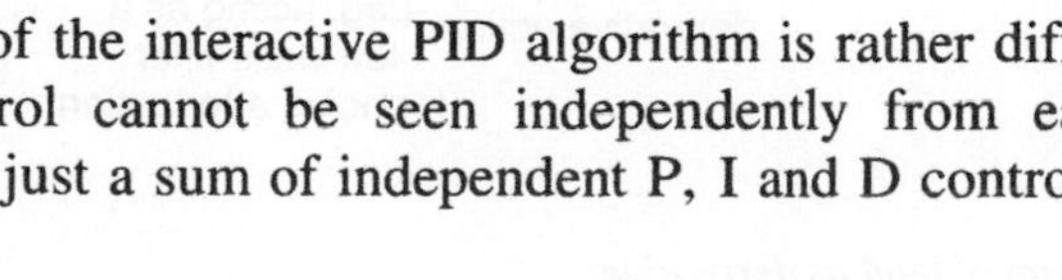

$$OP = K \times \frac{T_{DER} + 1}{\alpha T_{DER} + 1} \times \frac{1 + T_{INT}}{T_{INT}}$$

OP = Gain × Lead × PI-control

Figure 7.3
Real PID-controller

Since the interactive PID-controller makes use of a lead algorithm rather than using the classical mathematical derivative, it is best suited for real (field) process variables.

7.4.1   Real process variables (field originated)

A real process variable has electrical noise that come from field sensors or the connecting cables. It is therefore of great benefit that the PID algorithm has some noise reduction built in (Figure 7.4). The formula below represents an interactive PID algorithm:

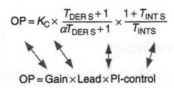

$$OP = K_C \times \frac{T_{DER}s + 1}{\alpha T_{DER}s + 1} \times \frac{1 + T_{INT}s}{T_{INT}s}$$

$$OP = Gain \times Lead \times PI\text{-}control$$

Figure 7.4
Real PID algorithm

7.5   Lead function – derivative control with filter

The following is an extract from Chapter 6 (Digital control principles) to remind us of the lead part acting as a *derivative function*. The field controller uses a lead algorithm for derivative control. The block diagram is shown in Figure 7.5.

7.5.1   Lead algorithm for derivative control (field or real PID controller)

$$\frac{Ts + 1}{\alpha Ts + 1} = \left(\frac{1}{\alpha Ts + 1}\right) \times (Ts + 1)$$

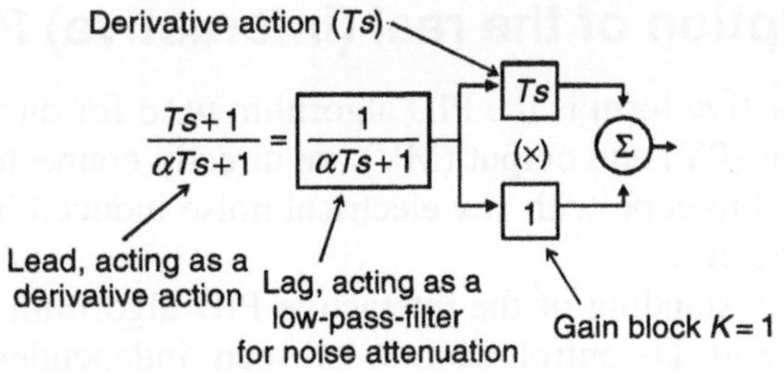

Derivative action (*Ts*)

$$\frac{Ts + 1}{\alpha Ts + 1} = \boxed{\frac{1}{\alpha Ts + 1}}$$
 → $\boxed{Ts}$, $\boxed{(\times)}$ → $\boxed{\Sigma}$ → , $\boxed{1}$

Lead, acting as a
derivative action   Lag, acting as a
low-pass-filter
for noise attenuation   Gain block *K* = 1

Figure 7.5
Block diagram of lead as derivative

7.6   Derivative action and effects of noise

The most important difference between non-interactive and interactive PID controllers is the different impact noise has on a controller's output. It must be remembered that derivative control multiplies noise.

7.6.1   Introduction to filter requirements

Both non-interactive PID and interactive PID controllers make use of a noise filter for process noise (*known as the process variable filter time constant TD*).

Since the derivative control of a non-interactive PID has no noise suppression of its own, noise will always be a major problem, even though a process variable filter may be used.

Since the derivative control of an interactive PID already has some noise suppression of its own, noise is not so much a problem, and is even less if a process variable filter is used.

It is recommended that a PV filter should be used in all cases where derivative control is being used. The author has observed numerous derivative control systems having excessive movement of the controller outputs due to the lack of PV filters. This type of problem is often incorrectly interpreted by personnel (in industrial plants) as being a problem of stability. Hence an important rule is: Make a clear distinction between noise and instability in industrial control applications.

As discussed earlier, noise and instability require treatment with different methodologies, as they are totally different problems.

Remember, a process variable filter, due to its lag action, reduces noise but may add to loop instability.

Exercise 5 (p. 242)

Introduction to Derivative Control

On the subject of D-control in non-interactive and interactive PID-controllers and the significance of noise.

7.7 Example of the KENT K90 controllers PID algorithms

$$\text{Proportional control} = K_1 = \frac{100}{\text{PB}}$$

$$\text{Integral control} = K_2 = \frac{100}{\text{PB}} \times \frac{\text{Scan period}}{\text{IAT(s)}}$$

$$\text{Derivative control} = K_3 = \frac{100}{\text{PB}} \times \frac{\text{DAT(s)}}{\text{Scan period}}$$

$$\text{Proportional} = K_1 \times Error$$

$$\text{Integral} = I_{n-1} + \frac{K_2(\text{SP}_n - \text{MV}_n)}{\text{MV range}}$$

$$\text{Derivative} = \frac{K_3(\text{MV}_{n-1} - \text{MV}_n)}{\text{MV range}}$$

$$\text{Result} = \text{proportional} + \text{integral} + \text{derivative}$$

$$= (K_1 \times \text{Error}) + \left(I_{n-1} + \frac{K_2(\text{SP}_n - \text{MV}_n)}{\text{MV range}} \right) + \left(\frac{K_3(\text{MV}_{n-1} - \text{MV}_n)}{\text{MV range}} \right)$$

$$= \frac{100}{\text{PB}} \left(E + \frac{1}{\text{IAT}} \int E \, dt + \text{DAT} \frac{d\text{PV}}{dt} \right)$$

8

Tuning of PID controllers in both open and closed loop control systems

8.1 Objectives

As a result of studying this chapter, and after having completed the relevant exercises, the student should be able to:

- Apply the procedures for open and closed loop tuning
- Calculate the tuning constants according to Ziegler and Nichols and according to Pessen
- Demonstrate how to perform fine tuning of closed loop control systems.

8.2 Objectives of tuning

There are often many and sometimes contradictory objectives, when tuning a controller in a closed loop control system. The following list contains the most important objectives for tuning a controller:

- *Minimization of the integral of the error*: The objective here is to keep the area enclosed by the two curves, the SP and PV trends, to a minimum. This is the aim of tuning, using the methods developed by Ziegler and Nichols as illustrated in Figure 8.1.
- *Minimization of the integral of the error squared*: As Figure 8.2 shows, it is possible to have a small area of error but an unacceptable deviation of PV from SP for a start time. In such cases special weight must be given to the magnitude of the deviation of PV from SP. Since the weight given is proportional to the magnitude of the deviation, the weight is multiplied by the error. This gives us error squared (error squared = error × weight). Many modern controllers with automatic and continuous tuning work on this basis.

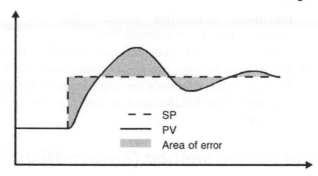

Figure 8.1
Integral on error

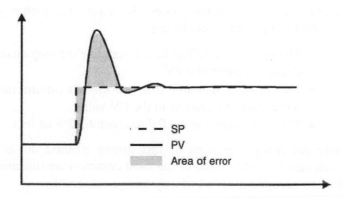

Figure 8.2
Integral on error square

- *Fast control*:   In most cases fast control is a principle requirement from an operational point of view; however, this is principally achieved by operating the controller with a high gain, quite often resulting in instability or prolonged settling times from the effects of process disturbances. Careful balances need to be obtained between the proportional or K_C function and the settings of the integral and particularly the derivative time constants T_{INT} and T_{DER} respectively.

- *Minimum wear and tear of controlled equipment*:   A valve or servo system for instance should not be moved unnecessarily frequently, fast or into extreme positions. In particular, the effects of noise, excessive process disturbances and unrealistically fast controls have to be considered here.

   Continual 'hunting' of the PV against the SP can result in a proportion of this, the magnitude depending on the controller gain, appearing on the controller's output. This, in many cases, can cause the driven actuator to 'vibrate' and this is quite often misconstrued as being caused by 'noise' when in fact it is caused by the gain of the controller, and as such the entire loop, being set too high in an attempt to 'speed-up' the response to the process (see Section 8.2.1).

- *No overshoot at start-up*:   The most critical time for overshoot is the time of start-up of a system. If we control an open tank, we do not want the tank to overflow as a result of overshoot of the level. More dramatically, if we have a closed tank, we do not want the tank to burst. Similar considerations exist everywhere, where danger of some sort exists. A situation of a tank having a maximum permissible pressure that may not be exceeded under any circumstances is an example here.

Note: Start-up is not the equivalent of a change of setpoint.

- *Minimizing the effect of known disturbances*:   If we can measure disturbances, we may have a chance to control these before the effect of them becomes apparent. See feedforward control for an example of an approach to this problem.

8.3   Reaction curve method (Ziegler–Nichols)

The reaction curve method of tuning relies on making a step change to the output of a controller and recording the process response. This method can be considered as an *open loop* approach, as the controller is *not* used in any way except for changing the OP value (in manual mode) to give the process the required step change to the MV.

The criteria we need to record are:

- The effective LAG or how long after the step change is made does a noticeable change occur in the PV
- The process reaction time or the maximum rate of change that occurs as represented by change in the PV value
- The time taken for the PV to reach 63.2% of its maximum value.

There are many variances of this tuning method, all utilizing the results from this reaction curve record. Three of the most common are discussed following the next section on how to generate a record of a systems reaction time.

8.3.1   The procedure to obtain an open loop reaction curve

Recording the PV response

Connect some form of recorder to the input (PV) signal to the controller. The recorder should ideally be capable of displaying two channels of information, the PV from the system into the controller, and the SP movement of the controller.

The record has to be plotted against a 0–100% PV vertical scale and a reasonably fast horizontal scale calibrated in minutes and fractions of minutes (not seconds). The vertical scale should be adjustable if using a paper strip recorder, so that the resultant change of the PV value covers a big a span as possible across the chart, this being required for measurement accuracy.

Controller mode

Place the controller in manual mode. This will ensure that we have an open loop in which the controller's action has no influence whatsoever when the PV value moves. This is because we are not interested in the controller's behavior, but only in the process's reaction characteristics.

Changing the process

When we make a step change to the output value of the controller, an appropriate reaction from the process will occur, appearing as a change in time of the PV value. This is the reaction characteristic of the process.

We must have enough process knowledge to know by how much we can change the output value of the controller without danger to the process itself.

Obtaining and analyzing the reaction curve

Observe the record of the reaction of the process. The plot we require is shown in Figure 8.3, where we can observe and measure the indicated parameters that are required to enable calculation of the P, I and D components of the controller, these being some or all of the ones listed below depending on, which analysis method you select to use.

- The point in time when the SP value was changed (the amount of this change is *not* important, it should be as large as possible as long as the process is not adversely effected by magnitude of the change).
- The time (in minutes and fractions of minutes) that elapses before a *noticeable* change is seen in the PV, this being measured as *L* or effective lag.
- The point of inflection (POI) on the PV curve.
- The point where the PV has changed by 63.2% (which is *not* necessarily the POI) to enable calculation of the LTC (loop time constant).

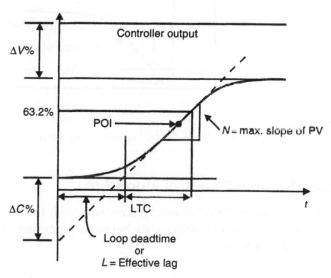

Figure 8.3
Ziegler–Nichols reaction curve

We cannot calculate the tuning constants before we have analyzed the curve using a few common sense considerations. The effective lag time (*L*) will be the principle effect and component of the integral time (T_{INT}) value. The slope, or rate of change of the process (*N*), will be the major factor influencing the controller gain setting, K_C, as it represents the gain or sensitivity of the process itself.

This leaves the derivative time constant to be determined (T_{DER}) and as this is introduced to correct the destabilizing effect of the integral action, a relationship between T_{DER} and T_{INT} must exist.

Ziegler and Nichols have derived formulas for optimum tuning, that takes into account, and relates the P, I and D values to each other. The optimum tuning obtained with these formulae is aimed at minimizing the integral of the error term (minimum area of error).

It does not take into account the magnitude of the error. Optimum tuning constants are invariably based on processes with a small deadtime and a first order lag.

As mentioned at the beginning of this section, there are three variations to this tuning method, Sections 8.4, 8.5 and 8.6 describe each of these in detail.

8.4   Ziegler–Nichols open loop tuning method (1)

From Figure 8.4, we have to derive a value for the effective lag (L), the time taken in decimal minutes until a noticeable rate of change is observed, and a value of N (the slope of the PV at the point of maximum rate of change).

From these two values we can calculate the tuning constants for P, PI and PID controllers according to the following Ziegler–Nichols formulae.

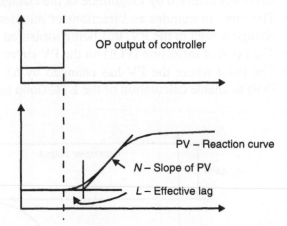

Figure 8.4
Ziegler–Nichols open loop tuning method (1) using rate of change (N) and effective lag (L) values

8.4.1   Ziegler–Nichols P control algorithm

Note that we obtain different tuning constants with the different combinations of control modes, and that a relationship exists between them that is echoed through the different modes are shown here.

$$\text{P control } K_{\text{C}} = \frac{\text{OP\%}}{\left(\dfrac{N\%}{\min \times L \min} \right)}$$

8.4.2   Ziegler–Nichols PI control algorithm

If we need to have integral action, the gain of the controller is reduced by 10% and the integral time constant, introduced to help eliminate the 'offset' value between the SP and PV in the ERR term, is set at three times the lag period (L in min). As the Integral output is summed with the proportional output contained within the controller gain, K_{C} can be reduced slightly, making the loop more stable. The loss in output resulting from this is gradually made up, in the integral time T_{INT}, by the integral action.

$$\text{PI control } K_{\text{C}} = 0.9 \times \frac{\text{OP\%}}{\left(\dfrac{N\%}{\min \times L \min} \right)}$$

$$T_{\text{INT}} = 3 \times L \text{ (min)}$$

8.4.3 Ziegler–Nichols PID control algorithm

Next, if we need to introduce some help in stabilizing the loop, we should introduce the derivative control. In doing this we see that the controller gain is increased by 20%. The integral time is made 33% faster (or shorter) and the derivative time constant is four times faster, or shorter, than the integral time. Put another way, the relationship between T_{INT} and T_{DER} is 4:1.

$$\text{PID control} \quad K_C = 1.2 \times \frac{OP\%}{\left(\dfrac{N\%}{\min \times L \, \min} \right)}$$

$$T_{INT} = 2 \times L \, (\min)$$
$$T_{DER} = 0.5 \times L \, (\min)$$

8.4.4 Examples of Ziegler–Nichols P, I and D open loop control algorithms

If we substitute the following values:

- OP% = DOP = 12.5%
- N = 35% per minute
- L = 0.65 min.

The settings for P, I and D can be summarized as follows:

MODE	K_C	T_{INT}	T_{DER}
P	$\dfrac{12.5}{35 \times 0.65} = \dfrac{12.5}{22.75} = 0.549$	–	–
PI	$0.9 \times \dfrac{12.5}{35 \times 0.65} = 0.9 \times \dfrac{12.5}{22.75} = 0.495$	$3 \times 0.65 = 1.95 \, (\min)$	–
PID	$1.2 \times \dfrac{12.5}{35 \times 0.65} = 1.2 \times \dfrac{12.5}{22.75} = 0.659$	$2 \times 0.65 = 1.3 \, (\min)$	$0.5 \times 0.65 = 0.325 \, (\min)$

8.5 Ziegler–Nichols open loop method (2) using POI

This version or method of deriving the gain, integral and derivative times uses the same response curve but which is made in a slightly different manner to the previous example. It is used where the process is controlled by a valve. To obtain the process curve, the following procedure is used:

- Bring the process to a desired setpoint on MANUAL control.
- Change the *valve position* a small amount, ΔV (%). The change should be large enough to produce a measurable response in the process, but not large enough to drive the process beyond normal operating range. A 5% valve change is a good starting point.
- Measure ΔC (%) and L on the process response curve.

The POI (point of inflection) is determined on the PV curve (point of maximum rate of change) and a tangential line is drawn through this, down through the horizontal axis and on until it crosses the vertical axis (Figure 8.5) (the time when the SP value was changed).

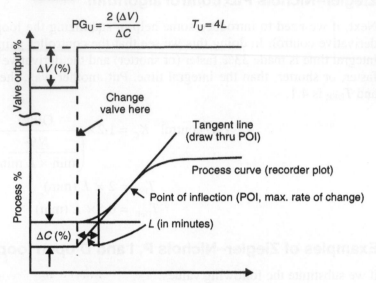

Calculate:

$$PG_U = \frac{2\,(\Delta V)}{\Delta C} \qquad T_U = 4L$$

Figure 8.5
Zeigler–Nichols open loop tuning method (2) using POI on the PV curve

From this can be calculated the following constants PG_U and T_U:

$$PG_U = \frac{2 \times (\Delta V)}{\Delta C} \quad \text{and} \quad T_U = 4L$$

Controller settings are determined from Table 8.1:

Controller	Proportional Only	Proportional Integral	Proportional Derivative (See Note Below)	Proportional Integral Derivative
Gain K_C	$0.5PG_U$	$0.45PG_U$	$0.71PG_U$	$0.6PG_U$
Integral time T_{INT}		$0.83T_U$		$0.5T_U$
Derivative time T_{DER}			$0.51T_U$	$0.125T_U$

Table 8.1
PID controller tuning parameter settings for open loop using Zeigler Nichols method

Note: The settings for a PD controller do *not* originate from the original Ziegler–Nichols paper.

It should be noted that a similar relationship of gain and integral/derivative times exists between this method and the previous one.

That is:

- The gain K_C in P mode = 0.5, in PI mode = 0.45 and PID = 0.6 or the gain ratios relate as 1 to 0.9 to 1.2
- In PID mode the ratio of T_{INT} to T_{DER} is again 4:1 (0.5 : 0.125).

Using this method, the slope or rate of change is quite often much easier to evaluate from a recorded chart.

8.6 Loop time constant (LTC) method

This method of tuning, as in the previous two examples, makes use of the reaction curve and is applicable when the system has a first order lag response as defined by a linear first order differential equation. This equation is expressed as:

$$\tau\frac{dc}{dt}+c=Kr \quad \text{or} \quad r \longrightarrow \boxed{\dfrac{K}{1+\tau\dfrac{d}{dt}}} \longrightarrow c$$

Where

c = output
r = input
K = gain
τ = time constant.

Inspection of a first order response curve will show that it is always falling off, i.e. the rate of response is at maximum in the very beginning and is continuously decreasing from that time onward. If the system continued to change at its maximum response rate, the rate that occurs at the origin, it would reach its final value (100%) in one time constant (T_{INT} time).

Figure 8.6 illustrates a first order curve derived from a step input. This curve gives numerical values to the change, and in the first period of time (in our case the integral time constant set by T_{INT}) the change equals 63.2%. In the second time period 63.2% of the remaining 36.8% will take place, and so on in every time interval. Theoretically the response never reaches 100%, but it does approach it asymptotically.

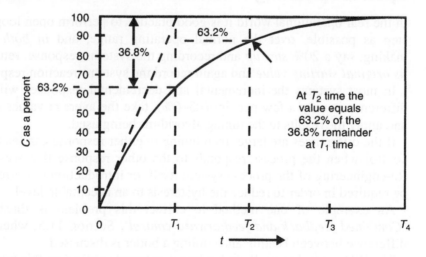

Figure 8.6
Response of a first order lag to a step input

By measuring both the loop deadtime and the loop time constant, the time from a noticeable change in the PV value to the time (in minutes) that a value of 63.2% is reached as shown in Figure 8.7, the following can be determined:

- $P_G = 1 / P_G$ (open loop)
- $I_G = LTC$
- $D_G = 0.25 \times I_G$.

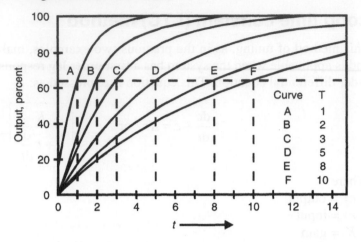

Figure 8.7
First order lag response curve

Exercise 7 (p. 252)

Open Loop Tuning Exercise

Provides practise in the reaction curve method of tuning.

8.7   Hysteresis problems that may be encountered in open loop tuning

In the real operational world it is good practice to perform open loop tuning with as big a step as possible, over the normal operating range, and *in both directions*, i.e. after making, say a 20% *step up* and recording the systems response, return the output *back to its original starting value* and again record the systems reaction response.

In most systems the incremental and decremental responses will be different. If this difference is only a few percent (<5–6%), take the average values of the two recordings and apply the results to the tuning algorithms being used.

If the differences are large, then tuning to either response can lead to instability or poor control when the process responds to the other response that was *not* used for tuning. Re-engineering of the process system itself, or introduction of corrective algorithms, will be required in order to reduce the hysteresis to an acceptable level.

An example of one method to correct this problem is illustrated in Chapter 11 '*Combined feedback and feedforward control*', Section 11.5, where correcting the time difference between heating and cooling a boiler is discussed.

The PID controller itself cannot be set or tuned to alleviate this type of problem.

8.8   Continuous cycling method (Ziegler–Nichols)

This method of tuning requires that we determine the critical value of controller gain (K_C) that will produce a continuous oscillation of a control loop. This will occur when the total loop gain (K_{LOOP}) is equal to one. The controller gain value (K_C) then becomes known as the ultimate gain (K_U).

Chapter 5, Sections 5.5 and 5.5.1 describe the requirements needed for a system to be considered stable.

We have to remember here that the loop is made up of several component parts, all of which contribute to the total gain of the loop (K_{LOOP}), and the only one that we can adjust is normally the controller's gain (K_C).

If we consider a basic liquid flow control loop consisting of:

1. *A measuring device*:   A venturi flow meter with a 4–20 mA output signal, fed to a controller
2. *A controller*:   A PID controller with 4–20 mA output signal, that is used to control an actuator
3. *A control device*:   A valve actuator which controls the flow rate of the process fluid and
4. The process.

When the product of the gains of all four of these component parts equals one, the system will become unstable when a process disturbance occurs (a setpoint change). It will oscillate at its natural frequency which is determined by the process lag and response time, and caused by the loop gain becoming one.

For example, if the system listed above, had the following gain characteristics:

Venturi gain   = 0.75
Control valve gain = 1.12
Process gain   = 0.98

then the *process* gain (as 'seen' by the controller) is calculated as:

$$0.75 \times 1.12 \times 0.98 = 0.8232.$$

With K_P equal to 0.8232, to make K_{LOOP} equal to 1, the value of K_C has to be

$$1/0.8232 = 1.215, \quad \text{giving} \quad K_{LOOP} = 0.75 \times 1.12 \times 0.98 \times 1.215 = 1$$

In order to observe the process dynamic characteristics only, we must not use any integral or derivative control during the process (as explained below) of determining the value of K_C in order to obtain a total loop gain, K_{LOOP}, equal to one (with no 'corrupting' phase shift introduced by the controller).

We can then measure the frequency of oscillation (the period of one cycle of oscillation), this being the ultimate period P_U.

In addition, we know that the final value of K_C is the critical gain of the controller (K_U). This gain value when multiplied with the unknown process gain(s), will give a loop gain, K_{LOOP}, of 1. From there we can stabilize the loop by reducing the value of K_C.

8.8.1   The stages of obtaining closed loop tuning (continuous cycling method)

1. *Put controller in P-control only*:   In order to avoid the controller influencing the assessment of the process dynamic, no integral or derivative control should be active. Make $T_{INT} = 999$ and $T_{DER} = 0$.
2. *Select the P-control to ERR = (SP − PV)*:   Make sure that P-control is working with PV changes as well as with SP changes. This enables us to make changes to the ERR term, and hence the controller output, by changing the SP value.
3. *Put the controller into automatic mode*:   We need a closed loop situation to obtain continuous cycling at the critical gain setting.

4. *Make a step change to the setpoint*:   To observe how the PV settles after a disturbance, change the SP value to simulate one. Before making this step change to the SP make sure the process is steady with only minor dynamic fluctuations visible.

5. *Actions based on the observation*:   If any oscillations that occur settle down quickly (or indeed there are no oscillation at all), then increase the value of K_C. The amount of increase to K_C depends on the rate and magnitude of change of the PV as a result of the last SP change.

   Then repeat 4 above, returning the setpoint back to its original value. When oscillations appear, and if they seem to be increasing in amplitude, terminate the exercise immediately and reduce the value of K_C to enable the process to stabilize. The total loop gain was >1, hence it amplified the SP change value. Repeat the exercise again, being more cautious with high values of K_C.

6. *Conclusion of tuning procedure*:   Once you obtain continuous cycling of the process, measure the cycle time and the value of K_C obtained for continuous cycling. This time is the *ultimate period* (P_U), and the value of K_C is the *ultimate gain* (K_U).

Reduce the value of K_C by 50% to stop the oscillations and return the SP to its original value to stabilize the process.

8.8.2   Calculation of tuning constants (continuous cycling method)

We will obtain different tuning constants with P, PI and PID control modes. However, your attention is drawn to the fact that the same relationships as discovered in the reaction curve method of tuning re-appear here. Controller settings are determined from Table 8.2.

Controller	Proportional Only	Proportional Integral	Proportional Integral Derivative
Gain K_C	$K_C = 0.5 \times K_U$	$K_C = 0.45 \times K_U$	$K_C = 0.6 \times K_U$
Integral time T_{INT}		$\dfrac{P_U}{1.2}$	$\dfrac{P_U}{2}$
Derivative time T_{DER}			$\dfrac{P_U}{8}$

Table 8.2
PID controller tuning parameter settings for closed loop using Zeigler Nichols continuous cycling method

Note: There are no values given for PD-control with this method, but the ratios used for open loop tuning can be applied if required.

Exercise 8 (p. 256)
Closed Loop Tuning Exercise
For practise in the techniques of closed loop tuning.

8.9 Damped cycling tuning method

This method is a variation of the continuous cycling method. It is used whenever continuous cycling imposes danger to the process, but a damped oscillation of some extent is acceptable. The steps of closed loop tuning (damped cycling method) are as follows:

8.9.1 Tuning method

1. *Put the controller into P-control only*: In order to avoid the controller influencing the assessment of the process dynamics, no I-control or D-control must be active.
2. *P-control on ERR = (SP − PV)*: Make sure that the P-control is working with PV changes as well as with SP changes. This enables us to make changes to the ERR term by changing the SP value.
3. *Put controller in automatic mode*: We need a closed loop situation to obtain damped cycling.
4. *Step change to the setpoint*: A step change to the SP causes a disturbance and we observe how the PV settles. Before making a step change to the SP, the process must be steady with only minor dynamic fluctuations visible.
5. *Actions based on the observation*: If any oscillations that occur settle down quickly (or indeed there are no oscillation at all), then increase the value of K_C.

The amount of increase to K_C depends on the rate and magnitude of change of the PV as a result of the last SP change.

Then repeat 4 above, returning the setpoint back to its original value. When oscillations appear, and if they seem to be increasing in amplitude, terminate the exercise immediately and reduce the value of K_C to enable the process to stabilize. The total loop gain was >1, hence it amplified the SP change value. Repeat the exercise again being more cautious with high values of K_C.

When a damped oscillation is obtained, as shown in Figure 8.8, note the value of K_C, this now being denoted as K_D. Then terminate the test by reducing the value of K_C. K_D is used to determine the gain later in this exercise.

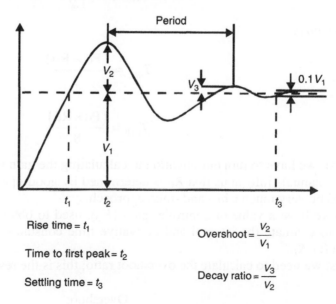

$$\text{Rise time} = t_1$$
$$\text{Time to first peak} = t_2$$
$$\text{Settling time} = t_3$$

$$\text{Overshoot} = \frac{V_2}{V_1}$$
$$\text{Decay ratio} = \frac{V_3}{V_2}$$

Figure 8.8
Damped oscillation decay ratio

8.9.2   Calculations

By measuring and dividing the amplitude of the first overshoot by the amplitude of the second overshoot the delay ratio P is found. The time (in minutes) between these two measured points gives a value for Pd (period of damping).

$$P = \text{Decay ratio} = \frac{\text{1st overshoot}}{\text{2nd overshoot}}$$

Then calculate the damping ratio δ from:

$$A = \frac{1}{2\Pi} \times \ln \times \left(\frac{1}{P}\right)$$

and then:

$$\delta = \frac{A}{\sqrt{1+A^2}}$$

In most cases the damping factor, δ, having a value of around 0.5 for a damped oscillation is acceptable.

We then need to evaluate Rd from P_U / P where P_U represents the ultimate period and P represents the actual period:

$$\frac{P_U}{P} = \sqrt{1-\delta^2} = \text{Rd}$$

This leads to the following formula for T_{INT} and T_{DER}:

PI-control:

$$T_{INT} = \frac{(\text{Pd} \times \text{Rd})}{1.2}$$

PID-control:

$$T_{INT} = \frac{(\text{Pd} \times \text{Rd})}{2}$$

$$T_{DER} = \frac{(\text{Pd} \times \text{Rd})}{8}$$

Next, we have to turn our attention to calculating the gain setting for the controller (K_C). Some manuals inform us that K_C is determined by good operator judgment; however this would be very much a hit- and -miss approach.

As we have a value of controller gain (K_D), used to obtain the damped cycle response used to evaluate the integral and derivative time constants. We can use this to obtain a value for K_U.

First we need to calculate the overshoot ratio; this is the result of

$$\frac{\text{Overshoot}}{\text{Steady-state change}}$$

We then calculate K_U from

$$K_U = \frac{K_D}{\text{Overshoot ratio}}$$

Achieving a value for K_U will let us use the Ziegler–Nichols closed loop formulas. These being

P-control:

$$K_C = 0.5 \times K_U$$

PI-control:

$$K_C = 0.45 \times K_U \text{ and}$$

PID-control:

$$K_C = 0.6 \times K_U$$

Figure 8.9 gives a graphical representation to obtain the damping ratio directly from the % overshoot that occurred in the PV as a result of a step change made to the controller output.

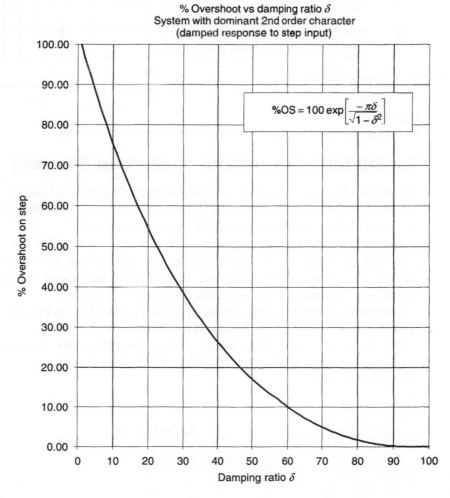

Figure 8.9
% Overshoot vs damping ratio system with dominant 2nd order character (damped response to step input)

8.9.3   Step responses

Figure 8.10 is used to determine the ultimate period (P_U) from the damped cycle period (P) (Courtesy of J.P. Stiekema).

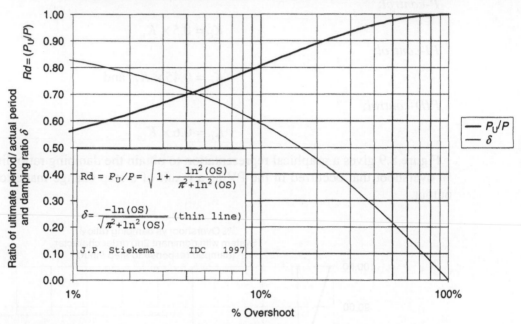

Figure 8.10
Damped step response of system with a dominant 2nd order characteristic

8.10   Tuning for no overshoot on start-up (Pessen)

This method is a variation of the continuous cycling method and it is used whenever no overshoot is permitted, even in the extreme case of start-up of the process. With start-up, we mean the transition from manual to automatic control.

An extreme start-up situation exists, if the setpoint and PV are very different when changing from manual to automatic control. In contrast to a change of setpoint, the change from manual to automatic control does not cause a step change in ERR. Therefore, the change does not directly affect P or D-control.

Examples for applying this tuning procedure, according to Pessen, is a closed tank that could burst or an open tank that could overflow.

The steps of closed loop tuning for no overshoot are the same as the ones for continuous cycling method (refer Section 8.8.1). The formulas developed for this case by Pessen are as follows:

PID-control:

$$K_C = 0.2 \times K_U$$

$$T_{INT} = \frac{P_U}{3}$$

$$T_{DER} = \frac{P_U}{2}$$

8.11 Tuning for some overshoot on start-up (Pessen)

This method is a variation of the continuous cycling method. It is used whenever no overshoot during normal modulating control is desired, but some overshoot at start-up is acceptable.

The steps of closed loop tuning for some overshoot are the same as the ones for continuous cycling method (refer Section. 8.8.1). The formulae developed for this case by Pessen are as follows:

8.11.1 The tuning constants

PID-control:

$$K_{\mathrm{C}} = 0.33 \times K_{\mathrm{U}}$$

$$T_{\mathrm{INT}} = \frac{P_{\mathrm{U}}}{2}$$

$$T_{\mathrm{DER}} = \frac{P_{\mathrm{U}}}{3}$$

8.12 Summary of important closed loop tuning algorithms

Tuning for	Continuous Oscillation	Pessen Some Overshoot	Pessen No Overshoot
K_{C}	$0.6 \times K_{\mathrm{U}}$	$0.33 \times K_{\mathrm{U}}$	$0.2 \times K_{\mathrm{U}}$
T_{INT}	$\dfrac{P_{\mathrm{U}}}{2}$	$\dfrac{P_{\mathrm{U}}}{2}$	$\dfrac{P_{\mathrm{U}}}{3}$
T_{DER}	$\dfrac{P_{\mathrm{U}}}{8}$	$\dfrac{P_{\mathrm{U}}}{3}$	$\dfrac{P_{\mathrm{U}}}{2}$

Table 8.3
Summary of closed loop PID controller tuning parameter settings for different controller responses

8.13 PID equations: dependent and independent gains

The general PID equation as applicable to digital (PLC) systems is the sum of four terms:

OP = Proportional + integral + derivative + bias (MANUAL) value

This equation can be represented in two ways, ISA (Instrument Society of America) (dependant gains) and independent gains.

In the independent gains equation, as the name suggests, all three PID terms operate independently. In the ISA equation a change in the proportional term also effects the integral and derivative terms (Figure 8.11).

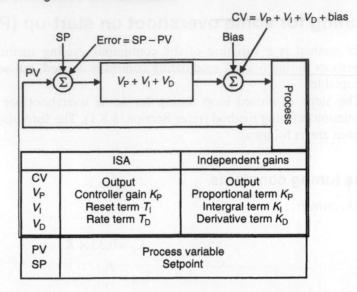

Figure 8.11
Closed loop control showing terms and comparison between ISA and independent gains equations

8.13.1   ISA equation

The ISA equation is interactive, that is, it contains dependent terms that mean if the controller gain K_C is changed, the integral and derivative terms also change.

$$CV = K_C \left| E + \frac{1}{T_{INT}} \int_0^t E \mathrm{d}t + \frac{T_{DER} \left[E - E(n-1) \right]}{\mathrm{d}t} \right| + \text{bias (MANUAL)}$$

or

$$CV = K_C \left| E + \frac{1}{T_{INT}} \int_0^t E \mathrm{d}t + \frac{T_{DER} \left[PV - PV(n-1) \right]}{\mathrm{d}t} \right| + \text{bias (MANUAL)}$$

Where

CV	= Output
K_C	= Controller gain constant (unitless)
T_{INT}	= Integral time constant (minutes per repeat)
T_{DER}	= Derivative time constant (minutes)
$\mathrm{d}t$	= Time between samples (minutes)
Bias	= Feedforward or output bias
E	= Error = to PV – SP or SP – PV
PV	= Process variable
PV$(n-1)$	= PV value from last sample
$E(n-1)$	= Error value from last sample.

8.13.2   Independent gains equation

This equation is non-interactive. As such P, I and D terms are adjusted independently.

$$CV = K_P E + K_I \int_0^t E\,\mathrm{d}t + \frac{K_D \left[E - E(n-1) \right]}{\mathrm{d}t} + \text{bias (MANUAL)}$$

or

$$CV = K_P E + K_I \int_0^t E\,\mathrm{d}t + \frac{K_D \left[PV - PV(n-1) \right]}{\mathrm{d}t} + \text{bias (MANUAL)}$$

Where

CV	= Output
K_P	= Proportional gain constant (unitless)
K_I	= Integral gain constant (1/sec)
K_D	= Derivative gain constant (seconds)
$\mathrm{d}t$	= Time between samples (seconds)
Bias	= Feedforward or output bias
E	= Error = to PV – SP or SP – PV
PV	= Process variable
$PV(n-1)$ =	PV value from last sample
$E(n-1)$	= Error value from last sample.

The ISA and independent gains constants can be compared as follows:

ISA Constants	Independent Gains Constants
Controller gain K_C (dimensionless)	Proportional gain K_P (dimensionless)
Reset term T_{INT} (minutes per repeat)	Integral gain K_I (inverse seconds)
Rate term T_{DER} (minutes)	Derivative term K_D (seconds)

To convert from ISA terms to independent gain terms:

$$K_P = K_C \quad \text{Unitless}$$

$$K_I = \frac{K_C}{T_{INT} \times 60\,\text{s/min}}$$

$$K_D = K_C (T_D)\, 60 \text{ s}$$

1: ISA Dependant Gains	0: AB Independent Gains
Setpoint	(Scaled) Setpoint
Proportional gain (K_C) (0.01)	Proportional gain (K_P) (0.01)
Reset time (T_1) (0.01 min/repeat)	Integral gain (K_I) (0.001/s)
Derivative rate (T_2) (0.01 min)	Derivative gain (K_D) (0.01 s)
Loop update time (0.01 s)	Loop update time (0.01 s)

Derivative Error   0:PV 1: Error:

Example

$$K_C = 2.2$$
$$T_{INT} = 0.8 \text{ min}$$
$$T_{DER} = 0.2 \text{ min}$$

$K_C = 220$ $K_P = K_C = 220$

$T_I = 80$ $K_I = \dfrac{K_C}{60T_I} = \dfrac{2.2}{60 \times 0.8} \times 1000 = 45.8$

$T_D = 20$ $K_D = K_C(T_D) \times 60 = 2.2 \times 0.2 \times 60 \times 100 = 2640$

Proportional band applications PB%:

$$PB\% = \frac{100}{K_C} = \frac{100}{2.2} = 45.5\%$$

9

Controller output modes, operating equations and cascade control

9.1   Objectives

As a result of studying this chapter, and after having completed the relevant exercises, the student should be able to:

- Demonstrate a clear understanding of controllers with multiple and independent outputs
- Clearly distinguish between saturation and non-saturation output limits
- Describe the concept and strategy of cascade control
- Select, and apply correctly, the controller options of initialization, PV-tracking and type of control equation
- Describe the concept of cascade control with multiple secondaries
- Demonstrate how to tune all controllers within a cascade control system.

9.2   Controller output

In order to enable controllers to be cascaded together certain system design requirements have to be made available. The most important one centers on the output section of a controller, in particular the primary one. Figure 9.1 shows a typical output section, or block, of a PID controller, illustrating the control signals and actions they perform upon the final output value. The functions of each of these will be discussed in this chapter.

9.2.1   Single or stand-alone controller output

The value of the final output of a single or stand-alone controller is affected by one of the two possible signals:

1. The first one is derived from the MANUAL mode, where a set or static value can manually be placed in the output, this value being considered a 'live zero' The controller itself has no knowledge of what this value is, and it can be anywhere for 0 to 100% of the output range.
2. The second one is when the controller is in AUTO mode and the PID actions now start to increment or decrement the MANUAL value in each scan time of the system.

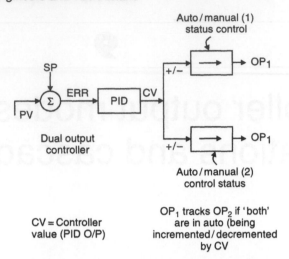

Figure 9.1
Dual output controller

9.3   Multiple controller outputs

A close inspection of Figure 9.1 shows that this controller can have two or more output blocks, all identical, but totally isolated from each other.

If we consider a controller with two (or more) output blocks, which result in final output signals OP_1, OP_2 to OP_N there exists many permutations of possible actions that this type configuration can perform, the most important being listed below:

9.3.1   Multiple controller output configurations

As each output block is independent of all the others that are attached to a controller, their absolute output values can be, and usually are, different from each other. Although the PID controller's action is continually reacting to the SP and PV values on its input, the results of the PID calculations will only affect, $(+ /0 /-)$, a particular output if the mode of that output is set to auto or cascade.

Figure 9.2 illustrates the output control strategy of a single or multi-output controller.

Assume initially, the Output of the Controller-1 (OP1) is set to manual mode or initialized and the Controller output (OP1) is cascaded or connected to another controller (see later in this chapter); and the Controller-2 is in auto mode. The Output of controller-2 (OP2) will be responding to the PID change requirements, but the Controller-1 Output (OP1) will be static at its manual value or initialized value. Only when Controller-1 (OP1) is set to AUTO mode or CASCADE mode, by either the MANUAL or INITIALISE control signals changing will OP1 (Output of Controller-1) then starts to respond to the PID commands.

The OP_1 value will then 'Track' the value of OP_2, although they may well have different absolute values. If the PID summing result says 'Increment' by a value of P_N in one scan time, then OP_1 will increase its value from OP_{1N} to $OP_{1N} + P_N$ and OP_2 will increase its output value by OP_{2N} to $OP_{2N} + P_N$, i.e. both outputs will change by the same magnitude, but maintain the differential value between them.

9.3.2   Limits of controller outputs

The controller itself has no knowledge of its final and absolute output value(s) from its output blocks. The result is that these outputs can be driven into saturation at 0% or 100% with the controller still trying to 'drive' them further below zero or above full scale.

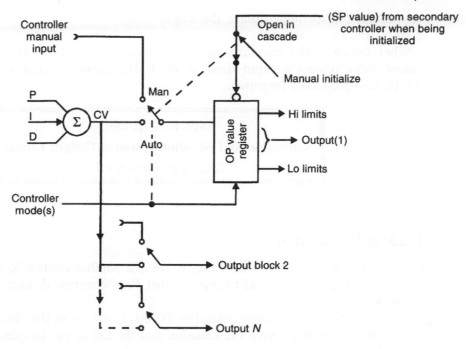

Figure 9.2
Output control, block control and interconnections

As we may well not know or be aware of this happening, and how and when and with what accuracy the outputs recover back into their operational range, we must be able to 'select' our requirements for the type of output limit calculations we need and why.

9.4   Saturation and non-saturation of output limits

There are two principle types of output limit calculations, the first, with output limits, allows saturation of the output based on P and D-control. The second does not allow output saturation to occur under any circumstances.

9.4.1   Saturation of the output

If the output of a controller is allowed to saturate, it allows it in the following manner:

- The controller calculates a VIRTUAL output value independent of any output-limit. These may be values far above 100% or far below 0%.
- Only the real output, which is the displayed output value, is limited by pre-defined output-limits.
- The real output then awaits the return of the virtual output to within the defined output-limits.
- Then, within the range of the output-limits, the real output follows the virtual output value exactly.
- Controllers driving field output normally use this kind of output-limit handling.

9.4.2   Non-saturation output limit calculation

Non-saturation of the output is achieved by ensuring that only the real output values are used for the calculation. If a single calculation results in an output value attempting to go beyond the pre-set output-limits, the output value will be set to the value of the output-limit it would have violated.

When the controller calculates the output value next time (in the next *scan*), the *real* output value (output = output limit) is used. The previous calculation, beyond output limits, has been totally forgotten.

Exercise 9 (p. 260)

Saturation and Non-Saturation of Output Limits

Illustrates Section 9.4.

9.5   Cascade control

Using the example of our feedheater, if we add another control loop, which is just to control the fuel flow, we will keep the fuel flow constant despite fuel flow pressure changes.

If the OP of the temperature controller TC drives the SP of this newly added fuel flow controller, FC, then we have the situation that the OP of the temperature controller TC then drives the *true flow* and *not* just a *valve position*.

Fuel flow pressure would practically no longer have any effect on the outlet temperature. This concept is called *cascade control*. The principle is shown in Figure 9.3.

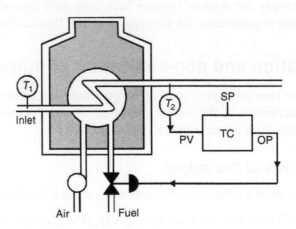

Figure 9.3
Single loop temperature control

9.5.1   Cascade control terms

In order to help identify which controller is which within a cascaded system, the following terms apply:

- The controller, whose SP *is driven by another controller's* OP may be called a 'downstream controller' (slave), or perhaps more often it is referred to as a '*Secondary controller*'.
- The controller whose OP *drives the* SP *of a secondary controller* is called an 'upstream controller' or '*Primary controller*' (master).

Multiple cascaded configurations

If we have more than two controllers in a cascade system,

- The highest upstream controller is called the *ultimate primary*.
- The lowest downstream controller is called the *ultimate secondary controller*.

If we examine the requirements needed to start-up such a feed heater cascade control system manually, it will give us insight into the principles of operation. This, in turn, will also give us the background required to understand PV-tracking, initialization and the different *equation types* used in the related control algorithms.

Referring to Figure 9.4, this illustrates the basic cascaded system, with the temperature controller (TC) being the primary control, and the fuel control (FC) the secondary controller.

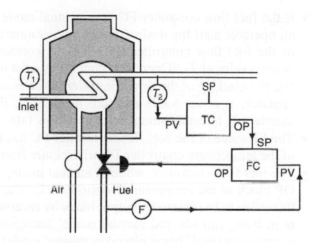

Figure 9.4
Two-controller basic cascade control

9.5.2   The concept of process variable or PV-tracking

PV-tracking is active if the secondary (FC) controller is *in manual mode*. Controllers can be set up to make use of PV-tracking or not. *It is the choice of the system designer.*

The concept is that an operator sets the OP value of the fuel controller manually until they find an appropriate value for the process. We assume that this output value is the optimum value for the process, that is we have set the fuel flow rate to a manual value that is correct to maintain the output temperature, T_2, at the required value.

This leads to the conclusion that no correction of the OP value is necessary *at this time*.

As no change of OP is the ideal requirement, no error (ERR) is permitted. To achieve this we have to keep the SP equal to the PV by the operator manipulating the OP value manually.

Hence for PV-tracking in MANUAL mode only: SP = PV. This is called PV-tracking.

The moment we change the mode to AUTOMATIC, the SP stays at its last value and is the reference previously created by manual manipulation of OP (when the mode was set to MANUAL).

The output of the flow controller (FC) has an ABSOLUTE value as determined by what was set by the operator at the time of the transition from manual to automatic sufficient to hold the fuel valve in its required position.

9.6 Initialization of a cascade system

Initialization is actually a kind of manual mode, in which the operator doesn't drive the OP value of the primary controller (our temperature controller, TC, in this case). Instead, our FC supplies its setpoint (SP) value, back up the cascade chain, to the OP of the controller that will be driving it (the FC's SP) when the system is in automatic mode. If selected, PV-tracking can take place in the primary controller as it would occur in normal manual mode.

9.6.1 Steps of initialization

Let us analyze how initialization is useful by looking into our feed heater example again (Figure 9.4).

- If the fuel flow controller FC is in manual mode and its OP value is driven by an operator until the desired outlet temperature (T_2) has been reached, the PV of the fuel flow controller (FC) has the correct value in order to obtain the desired value of T_2. (Open loop conditions exist in this loop at this point.)
- Via PV-tracking of the flow controller, its setpoint value, as manipulated by the operator, is at the value required to give the FC output its correct value to maintain T_2 by establishing the correct flow rate.
- The SP value of the fuel flow controller FC has the exact value that the output of the temperature controller TC will require from it.
- The fuel flow controller, while in manual mode, passes back its SP value to the OP block of the temperature controller TC. The temperature controller allows this value to be set into its output block by receiving a signal from the FC that it is in both 'cascade and manual mode' (accepting this command in a similar manner as to itself being placed in manual mode).
- This is called *Initialization*. If the primary controller (TC) performs PV-tracking, then the temperature SP follows the true temperature value, the PV of the primary controller.
- When the operator has found the correct OP value of FC, we have, by default, obtained an SP value for the correct flow. SP = PV.
- If we matched this SP with the primary controller's OP value, then we have the correct SP for temperature as well.
- All there is to do is to put the secondary controller into CASCADE mode and the associated primary controllers output block should switch automatically to AUTO mode.
- We have thus achieved a smooth (bumpless) transfer from manual to automatic control.

9.7 Equations relating to controller configurations

In cascade control, outputs from controllers drive the setpoints of secondary controllers, and, in essence all of these controllers consist of, or are capable of, independently calculating P, I and D algorithms, based on the error value derived from the two inputs, the setpoint value (SP) and the process variable PV. There are occasions where only certain functions within a cascade chain are required, and it becomes necessary to 're-arrange' the way the P, I and D functions are driven from the PV and SP variables.

There are three ways to do this, and they are known as *controller equations type A, B or C*. Equations A and C are the more commonly used of the three, and are interrelated, so these will be considered together.

Figure 9.5 illustrates the interconnections of a controller that determines the type of equation it represents.

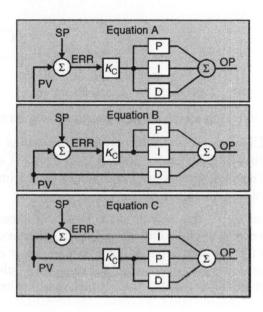

Figure 9.5
Equations types A, B and C

9.7.1   Equation type A

In Equation type A all control is based on the error term (ERR).

A controller using equation type A makes no distinction between a disturbance in the PV input and an operator action on the SP. This is the standard way of calculating control actions of a PID-controller and this has been the way in which we have considered all controllers so far in this book.

Where PV changes are fairly smooth with minimal or no step changes, they will not cause dramatic or sudden changes to the controller's output. Additionally the SP of the controller is normally never or very seldom moved again not causing rapid changes of output, but in contrast, an operator may drive a valve through its complete range by a large step change of the SP.

In such situations, we could consider the operator to be the most dangerous disturbance in the system.

Hence, when we require a smooth transition, even if the operator changes the SP dramatically, we need to 're-arrange' the construction of the controller to help us achieve this requirement. This leads to equation type C.

9.7.2   Equation type C

As can be seen from Figure 9.5, Equation B works as PI controller on ERROR (ERR – PV – SP) and works as a D-controller on the PV only. Equation type C configures the controller so that we can eliminate the problem of step changes to the output occurring by large and rapid changes being made to the setpoint value by the operator. We must remember that in most systems the SP of a primary controller is seldom changed much, either in magnitude or time. However when the SP is changed we need to ensure the resultant output change is as soft and gentle as possible, particularly when it is driving SPs of secondary controllers. A nice way to do this is to integrate the step change, illustrated as:

Resultant controller output step change by a SP change

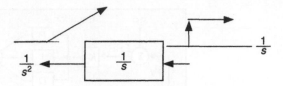

Referring to Figure 9.5 we see how equation C differs from the normal equation type A by:

- The proportional and derivative functions operating, via the gain block K_C, directly from the PV and *not* from the ERR term.
- The ERR term is used exclusively by the integral function, again derived from either PV – SP or SP – PV.

This means that a step change made to the SP becomes an integrated (ramped) output from the controller. This kind of control calculation calculates an identical PID-control action as with equation type A *if the setpoint is a constant*. This maintains the same control against real disturbances and the same loop behavior.

The SP is the only variable treated differently. The details of equation type C are as follows:

Equation C and the P-control

When the SP = constant – What is the difference?

$$\Delta\, OP = K \times \Delta(PV - SP) \qquad \text{[Eq-type A]}$$
$$\Delta\, OP = K \times \Delta\, PV \qquad \text{[Eq-type A]}$$

Answer: No difference, provided SP = constant.
Notes:

- Observe that the change of ERR, where ERR = $\Delta(PV - SP)$, is identical to the term $\Delta\, PV$ (the change of PV).
- There is identical P-control action based on PV, but the SP is ignored totally.
- The SP is not even part of the formula any more.
- The operator can do what he/she wants with the SP, but this will have no influence on P-control if equation type C is active.

Equation C and the I-control

The availability of an integral control is the reason for the existence of these controller equations because:

- There are no differences in I-control with different equation types.
- I-control is available to the operator at all times for smooth bumpless changes of control from one SP to another.
- I-control will never cause any bump if a SP change takes place.
- Since the SP is an outside influence as far as the loop is concerned, the integral on SP has no effect on loop stability.

Equation C and the D-control

When the SP = constant – What is the difference ?

$$\Delta\,OP = K \times T_{DER} \times \Delta(\Delta(PV - SP)) \qquad \text{[Eq-type A]}$$

$$\Delta\,OP = K \times T_{DER} \times \Delta(\Delta(PV)) \qquad \text{[Eq-type C]}$$

Answer: No difference, provided SP = constant.

Notes

- Observe $\Delta(\Delta(SP - PV)$ is identical to $\Delta(\Delta(PV)$ if there is no change of SP.
- There is identical D-control action based on PV, but the SP is ignored totally.
- The SP is not even part of the formula any more.
- The operator can do what he wants with the SP as he has no influence on D-control if equation type C is active.

This is one more example of the use and benefit of incremental algorithms.

9.8 Application notes on the use of equation types

We have to make a careful assessment of the process and the process strategy before we decide on a particular equation type. As a general rule we can use the different types as follows:

9.8.1 Application of equation type A

This is a general purpose calculation to be used if no special reason exists to use another type.

Important note: Eq-type A is a must for secondary controllers. If eq-type C were to be used in a secondary controller, I-control would be the only control between the OP of the primary controller and the OP of the secondary controller. This would add an unnecessary phase lag of 90 for the primary controller's loop. The result could be an unnecessary destabilization of the primary loop of a cascade control system.

9.8.2 Application of equation type B

The principle considerations, how the control algorithms can work on PV only, are the same as explained for equation type C.

Equation type B works as a PI-controller on error (ERR = PV – SP) and works as a D-controller on PV only.

Since eq-type B is in between eq-type A and eq-type C, it is thus left to the discretion of the user to make decisions about the use of eq-type B. If, for example, a secondary controller needs D-control for stability of the secondary loop and the OP of a primary controller contains all the control actions required for the primary loop (including D-control), then the secondary controller may be best used with eq-type B. In such a case, the full control action of the primary controller is passed on to the OP of the secondary controller via control of the secondary.

9.8.3 Application of equation type C

Equation type C works as I-controller on error (ERR = PV – SP) and works as a PD-controller on PV only.

Mainly used as the ultimate primary controller. An operator cannot cause any sudden control actions that would result in sudden extreme positions of valves and other control equipment. This can only be fully appreciated if one has heard the noise created by the sudden hitting of an extreme valve position of a large valve. It can be felt in almost the whole plant as a big bang. This is most decidedly not good for maximizing the life of a valve.

Equation Type	P Mode	I Mode	D Mode	Comments on Use
A	PV – SP SP – PV	PV – SP SP – PV	PV – SP SP – PV	Standard controller
B	PV – SP SP – PV	PV – SP SP – PV	PV PV	Special uses
C	PV PV	PV – SP SP – PV	PV PV	Primary controller

Table 9.1
Comparison of PID equation types and error calculations

9.9   Tuning of a cascade control loop

The approach for tuning is fairly straightforward. Firstly, tune the ultimate secondary controller (the most downstream controller) which in our example is the FC controller. Then considering that controller as part of the process, tune the next upstream controller (the one whose output drives the setpoint of the last tuned controller). Continue in this manner, remembering to consider the last tuned controller as part of the process loop, finally tuning the most primary controller, again in our example the TC controller.

9.9.1   Secondary controller

Secondary controllers are mostly used as flow controllers. Flow loops are in most cases intrinsically stable. Therefore, no D-control is required and most flow controllers are PI-controllers. Tuning is done with due consideration given to a sufficiently good control response and minimum wear and tear of the valve. The value of K should be smaller than 1 in order to pass on the full range of the primary controller's OP to the OP of the secondary controller.

9.9.2   Primary controller

Primary controllers normally control a dynamically more complex loop and require careful stability considerations. Our example of a feed heater shows clearly that the temperature controller TC has to cope with most of the process lag. In most cases, primary controllers are therefore PID-controllers.

> **Exercise 11 (p. 267)**
>
> **Cascade control**
>
> Will give practical experience on the topics of cascade control.

9.10 Cascade control with multiple secondaries

A control strategy can include controllers with multiple output calculations. In most cases, controllers with multiple outputs are primary controllers in a cascade control system with more than one secondary controller.

9.10.1 Multiple output calculations

The result of the primary controller's PID calculation is the controller's dynamic output. In digital controllers, this is the calculated value CV, which is calculated for each scan time interval and is used to increment each output independently.

As every output of a controller may have a different absolute value at any given time, every output is incremented individually.

In actual fact, each output is calculated independently of each other with independent initialization, limit and alarm handling. As the amount of data for multiple outputs is too much for one display, industrial control equipment will display only the first output in the main detail display and has subsequent displays for additional data like multiple outputs. One has to be aware of this from an engineering point of view in order to define the most significant output to be the first output of a controller. From an operator's point of view, it is important to know that the most prominently displayed output value may not be the only output to be monitored, it may just be the most important one.

Exercise 12 (p. 271)

Cascade control with one primary and two secondaries

Will give practical experience on the topics of controllers with multiple outputs.

10

Concepts and applications of feedforward control

10.1 Objectives

As a result of studying this chapter and after having completed the relevant exercises, the student should be able to:

- Describe the concept and strategy of feedforward control
- Develop and then clearly describe the tuning procedures for feedforward control.

10.2 Application and definition of feedforward control

If, within a process control's feedback system, large and random changes to either the PV *or lag time* of the process occur, the feedback action becomes very ineffective in trying to correct these excessive variances. These variances usually drive the process well outside its area of operation, and the feedback controller has little chance of making an accurate or rapid correction back to the SP term.

The result of this is that the accuracy and standard of the process becomes unacceptable. Feedforward control is used to detect and correct these disturbances before they have a chance to enter and upset the closed or feedback loop characteristics.

It must be remembered that feedforward control does not take the process variable into account; it reacts to sensed or measurement of known or suspected process disturbances, making it a compensating and matching control to make the impact of the disturbance and feedback control equal.

The difference between feedforward and feedback control can be considered as:

- Feedforward is primarily designed and used to prevent errors (process disturbances) entering or disturbing a control loop within a process system.
- Feedback is used to correct errors, caused by process disturbances, that are detected within a closed loop control system. These errors can be foreseen and corrected by feedforward control, prior to them upsetting the control loop parameters.

It is this second factor alone that makes feedforward a very attractive concept. Unfortunately, for it to operate safely and efficiently, a sound knowledge is required both of the process and the nature of all relevant disturbances.

10.3 Manual feedforward control

Feedforward is a totally different concept to feedback control. A manual example of feedforward control is illustrated in Figure 10.1. Here, as a disturbance enters the process, it is detected and measured by the process operator. Based on their knowledge of the process, the operator then changes the manipulated variable by an amount that will minimize the effect of the measured disturbance on the system.

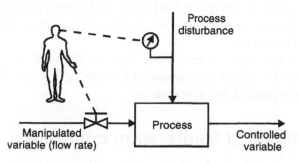

Figure 10.1
Manual feedforward control

This form of feedforward control relies heavily on the operator and their knowledge of the operation of the process. However, if the operator makes a mistake or is unable to anticipate a disturbance then the controlled variable will deviate from its desired value and if feedforward is the only control, an uncorrected error will exist.

10.4 Automatic feedforward control

Figure 10.2 illustrates the concept of automatic feedforward control. Disturbances that are about to enter a process are detected and measured. Feedforward controllers then change the value of their manipulated variables (outputs) based on these measurements as compared with their individual setpoint values.

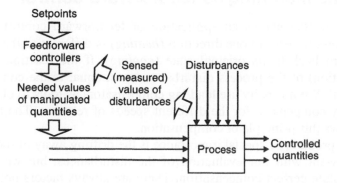

Figure 10.2
Automatic feedforward control

Feedforward controllers must be capable of making a whole range of calculations, from simple on–off action to very sophisticated equations. These calculations have to take into account all the exact effects that the disturbances will have on controlled variables.

Feedforward control, although a very attractive concept, places a high requirement on both the systems designer and the operator to both mathematically analyze and understand the effect of disturbances on the process in question.

As a result feedforward control is usually reserved for the more important or critical loops with a plant.

Pure feedforward control is rarely encountered, it is more common to find it embedded within a feedback loop where it assists the feedback controller function by minimizing the impact of excessive process disturbances.

In Chapter 11 (combined feedback and feedforward control), we will be examining the concepts and applications of feedforward control when combined with a cascaded feedback system.

It is important to remember that feedforward is primarily designed to reduce or eliminate the effect of changes in both process reaction times and the magnitude of any measured process variable change.

10.5   Examples of feedforward controllers

As discussed above, feedforward controllers can be required to carry out simple (on–off) control up to high order mathematical calculations. Due to the vast variances of requirements for feedforward controllers they can be considered as functional control blocks. They can range, as stated, from simple on–off control to lead/lag (derivative and integral functions) and timing blocks. Their range of functionality is virtually unlimited as most systems allow them to be 'programed' in as software-based math functions.

The four basic requirements for the composition of a feedforward controller are:

1. A recognizable *input* (derived from the measured disturbance)
2. A *setpoint* or point of origin and control
3. A math function operating on the *input* and *setpoint* values
4. An *output* which is the result of this math function.

In essence, the controller's action can be described purely by the mathematical function it performs.

10.6   Time matching as feedforward control

Figure 10.3 shows an application of feedforward control where the time taken for a process to react in one direction (*heating*) is different to the time taken for the process to return back to its original state (*cooling*). If the reaction curve (dynamic behavior of reaction) of the process disturbance is not equal to the control action, it has to be made equal. We normally use lead/lag compensators as tools to obtain equal dynamic behavior. They compensate for the different speeds of reaction. The block diagram in Figure 10.4 shows this principle of compensation.

A problem of special importance is the drifting away of the PV. We can be as careful as we want with our evaluation of the disturbances, but we never reach the situation of absolute perfect compensation. There are always factors not accounted for. This causes a drifting of the PV which has to be corrected manually from time to time, or an additional feedback control has to be added.

Figure 10.5 shows a carefully designed example, taking into account all major and measurable variables possible.

The feedforward control shown in Figure 10.5 uses mass flow calculation for the inlet flow and uses a fuel flow controller to avoid the influence of fuel flow pressure changes.

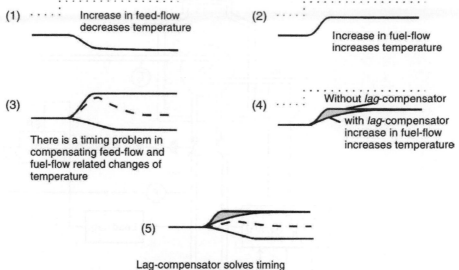

(1) Increase in feed-flow decreases temperature

(2) Increase in fuel-flow increases temperature

(3) There is a timing problem in compensating feed-flow and fuel-flow related changes of temperature

(4) Without *lag*-compensator with *lag*-compensator increase in fuel-flow increases temperature

(5) Lag-compensator solves timing problem in feed-flow-related changes of temperature

If the reaction curve for feed-flow related changes of temperature is faster than the reaction curve for fuel-flow changes, a lead-compensator would be required

Figure 10.3
Time matching of feedforward control

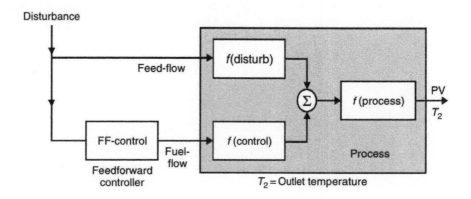

Figure 10.4
Block diagram of feedforward control

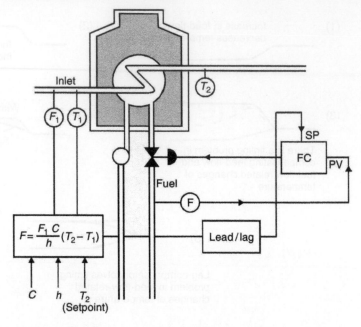

Figure 10.5
Feedforward control for feed heater

11

Combined feedback and feedforward control

11.1　Objectives

As a result of studying this chapter, and after having completed the relevant exercises, the student should be able to:

- Indicate the concept and strategy of combined feedback and feedforward control
- Demonstrate how to develop tuning procedures for combined feedback and feedforward control.

11.2　The feedforward concept

Chapter 10 illustrated the concepts of feedforward control and showed that one problem it gives us is drifting of the PV from the systems SP value. This is caused solely because the PV is *not* taken into account in feedforward control, if it was it would become a feedback (closed loop) controlled system.

Examination of the feedforward concept shows us that it is normally used to minimize the impact of disturbances on a process. This is achieved by detecting and measuring a process disturbance and changing a related manipulated variable before the disturbance has an adverse effect on the process itself.

It is important to remember that process disturbances constitute anything from unexpected changes in either: *magnitude* of pressure, flow, temperature and any other physical quantity associated with the process or *changes in time*, of any of the process responses.

This latter variable, *time*, is very often overlooked as a quantity that may need correcting in a process environment. This is illustrated in Figure 10.3 where we use feedforward to equalize the difference in heating and cooling times of a feed heater system.

This should make the process responses both in magnitude and time, the same, irrespective of the direction taken by the PV value. If this is achieved, tuning of the system is made easier, with the result the control is more stable and accurate.

11.3　The feedback concept

Chapter 5 explains the concepts of closed loop control and stability as related to feedback systems. In general terms it's accepted that a feedback system operates more accurately and efficiently if both process disturbances and time delays (lag times) are kept to the minimum. It then becomes apparent that feedforward control can be used to achieve this requirement being made by the feedback control.

If we then combine both an accurately configured feedforward system with a well-tuned feedback system the result should become almost an optimally operating control system.

11.4   Combining feedback and feedforward control

Figure 11.1 illustrates a concept where we combine both control methods into our feed heater control system.

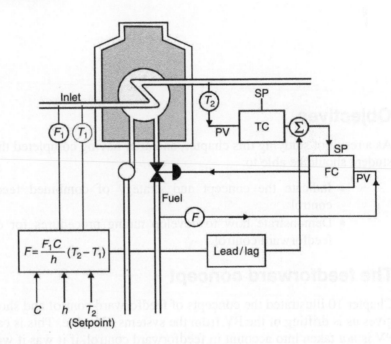

Figure 11.1
Combined feedback and feedforward control

Chapter 9 'Controller output modes, operating equations and cascade control' and Chapter 10 'Concepts and applications of feedforward control' cover most of the aspects which have to be considered when using feedback and feedforward control. Here we will concentrate on the impact of the *summer block* and some tuning aspects of the combined control system.

11.5   Feedback–feedforward summer

Referring to Figure 11.1 we see that we have only one manipulated variable, the fuel flow, but two control concepts combined.

The mass flow feedforward control, equating as

$$F = \left(\frac{(F_1 C)}{(h)} \right) (T_2 - T_1)$$

and the feedback cascaded control would appear to compete for the use of the one manipulated variable, the fuel flow.

However, if we remember the concept of compensation which governs feedforward control, the output of the feedforward control (from the lead/lag block) would usually be passed directly onto the SP of our fuel flow controller.

To this value coming from the feedforward control, we have to add the output value of the primary controller (temperature controller, TC). It is important to remember that these are incremental and *not* absolute calculations.

11.6 Initialization of a combined feedback and feedforward control system

As we have a combination of feedforward, feedback and cascade control the method of initialization is important.

The value for the *initialization of the* OP *of the primary controller (TC)* is calculated from the sum of:

- The value of the *feedforward signal* from the lead/lag block and (+)
- The SP *value* of the secondary controller *(FC)*.

As fluctuations occur to the inlet temperature and the inlet flow rate varies (F_1 and T_1 variables), and depending on which direction (up or down) they occur, the output of the lead / lag block will vary in accordance with the functional algorithm and lead/lag time constants which comprise the feedforward control.

The magnitude and rate of change of this signal has to be compatible with any of the F_1 and T_1 variances so that they have minimal or no influence on the outlet temperature T_2.

The summing block should be considered as part of the output block of the primary controller (TC).

11.7 Tuning aspects

Since strong feedback control has a tendency toward instability, it should be avoided if possible. Therefore, if the feedforward control is already doing the major part of the required control and feedback control is just there to eliminate drift of PV, proceed as follows:

- Tune the secondary controller used by feedback and feedforward control
- Tune the feedforward control
- Tune feedback control using the formulas developed by Ziegler and Nichols
- Evaluate the speed of drift of PV.

If the drift of the process variable is insignificant, reduce K of the primary controller using process knowledge and personal judgment.

Remember that in this case, the feedback control is just a supplement to feedforward control and must not introduce any form of oscillations or instability.

Exercise 13 (p. 276)

Combined feedforward and feedback control

Will give practical experience of combined feedback and feedforward control.

12

Long process deadtime in closed loop control and the Smith Predictor

12.1   Objectives

As a result of studying this chapter, and after having completed the relevant exercises, the student should be able to:

- Demonstrate the correct use of a process simulation for process variable prediction
- Show how control loops with long deadtimes are dealt with correctly
- List the procedures for tuning of control loops with long deadtimes.

12.2   Process deadtime

Overcoming the deadtime in a feedback control loop can present one of the most difficult problems to the designer of a control system. This is especially true if the deadtime is >20% of the total time taken for the PV to settle to its new value after a change to the SP value of a system.

We have seen that little or no deadtime in a control system presents us with a simple and easy set of algorithms that when applied correctly give us extremely stable loop characteristics.

Unfortunately, if the time from a change in the manipulated variable (controller output) and a detected change in the PV is excessive, any attempt to manipulate the process variable before the deadtime has elapsed will inevitably cause unstable operation of the control loop.

Figure 12.1 illustrates various deadtimes and their relationship to the PV reaction time.

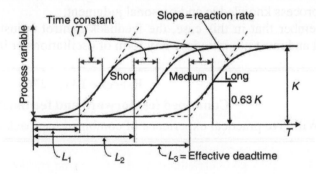

Figure 12.1
Reaction curves showing short, medium and long deadtimes

12.3 An example of process deadtime

Process deadtime occurs in virtually all types of process, as a result of the PV measurement being some distance away, both physically and in time, from the actuator that is driven by the manipulated variable.

An example of this is in the overland transportation of material from a loading hopper to a final process, this being some distance away. The critical part of the operation is to detect the amount of material arriving at the end of its journey, the end of the conveyor belt, and from this performing two functions:

1. To 'tell' the ongoing process how much material is arriving and
2. To adjust the hopper feed rate at the other end of the belt.

Figure 12.2 illustrates this problem: the controller is measuring the weight of arriving material that during its journey from the supply hopper has encountered some loss due to spillage from the conveyor. Also the amount of material deposited on the belt has varied due to variability of the amount, or head, of material in the hopper.

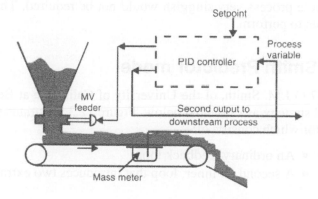

Figure 12.2
Illustration of a long conveyor system giving an excessive deadtime to the control loop

The deadtime can be calculated very simply by the product of the belt speed and the distance between the input hopper, where the action of the manipulated variable (controller output) occurs and the PV or point where the belt weigher is located.

In this example, the controller measures the weight/meter/minute of the arriving material, compares this with the SP and generates an output, but now it must wait for the deadtime period, which in this example is about 10 min, before seeing a result of this change in the value of the MV. If the controller expects a result before the deadtime has elapsed, and none occurs, it will assume that its last change had no effect and it will continue to increase its output until such time as the PV senses a change has occurred. By this time it will be too late, the controller will have overcompensated, either by now supplying too much or too little material.

The magnitude of this resultant error will depend on the sensitivity of the system and the difference between the assumed and actual deadtime. That is, if the system is highly sensitive (high gains and fast responses tuned into it) it will affect large movements of the inlet hopper for small PV changes. Also if the assumed deadtime is much shorter than the actual deadtime it will spend longer time changing its output (MV) before sensing a change in the PV.

12.3.1   Overcoming process deadtime

Solving these problems depends, to a great extent, on the operating requirement(s) of the process. The easiest solution is to 'detune' the controller to a slower response rate. The controller will then not overcompensate unless the deadtime is excessively long. The integrator (I mode) of the controller is very sensitive to 'deadtime' as during this period of inactivity of the PV (an ERR term is present) the integrator is busy 'ramping' the output value.

Ziegler and Nichols determined the best way to 'detune' a controller to handle a deadtime of D min is to reduce the integral time constant T_{INT} by a factor of D^2 and the proportional constant by a factor of D.

The derivative time constant T_{DER} is unaffected by deadtime as it only occurs after the PV starts to move.

If, however, we could inform the controller of the deadtime period, and give it the patience to wait and be content until the deadtime has passed, then detuning and making the whole process very sluggish would not be required. This is what the Smith Predictor attempts to perform.

12.4   The Smith Predictor model

In 1957 O.J.M. Smith, of the University of California at Berkeley proposed the predictor control strategy as explained below. Figure 12.3 illustrates the mathematical model of the predictor which consists of:

- An ordinary feedback loop
- A second, or inner, loop that introduces two extra terms into the feedback path.

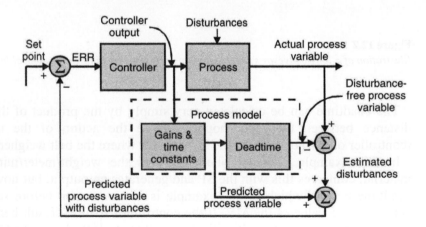

Figure 12.3
The Smith Predictor model

12.4.1   First term explanation (disturbance-free PV)

The first term is an *estimate* of what the PV would be like in the absence of any process disturbances. It is produced by running the controller output through a model that is designed to accurately represent the behavior of the process without taking any load disturbances into account. This model consists of two elements connected in series.

1. The first represents all of the process behavior not attributable to deadtime. This is usually calculated as an ordinary differential or difference equation that includes estimates of all the process gains and time constants.
2. The second represents nothing but the deadtime and consists simply of a time delay; what goes in, comes out later, unchanged.

12.4.2   Second term explanation (predicted PV)

The second term introduced into the feedback path is an estimate of what the PV would look like in the absence of both disturbances and deadtime. It is generated by running the controller output through the first element of the model (gains and TCs) but *not* through the time delay element.

It thus *predicts* what the disturbance-free PV will be like once the deadtime has elapsed.

12.5   The Smith Predictor in theoretical use

Figure 12.4 shows the Smith Predictor in a practical configuration, or as it is really used.

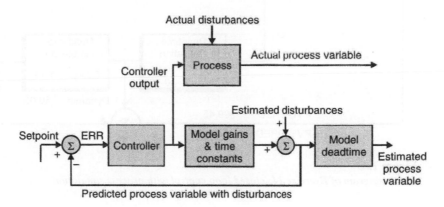

Figure 12.4
The Smith Predictor in use

It shows an estimate of the PV (with both disturbances and deadtime) generated by adding the estimated disturbances back into the disturbance-free PV. The result is a feedback control system with the deadtime outside the loop.

The Smith Predictor essentially works to control the modified feedback variable (the predicted PV with disturbances included) rather than the actual PV.

If it is successful in doing so, and the process model accurately emulates the process itself, then the controller will simultaneously drive the actual PV toward the SP value, irrespective of SP changes or load disturbances.

12.6   The Smith Predictor in reality

In reality there is plenty of room for errors to creep into this 'predictive ideal'. The slightest mismatch between the process dynamic values and the model can cause the controller to generate an output that successfully manipulates the modified feedback variable but drives the actual PV into nihilism, never to return.

There are many variations on the Smith Predictor principle, but re-admits, especially long ones remain a particularly difficult control problem to control and solve.

12.7   An exercise in deadtime compensation

We have seen that if a long deadtime is part of the process behavior, the quality of control becomes unacceptably low. The main problem lies in the fact that the reaction to an MV change is not seen by the PV until the deadtime has expired. During this time, neither a human operator nor an automatic controller knows how the MV change has effected the process.

Exercise 14 (p. 279) illustrates the concepts of deadtime compensation, based on the arrangement shown in Figure 12.5.

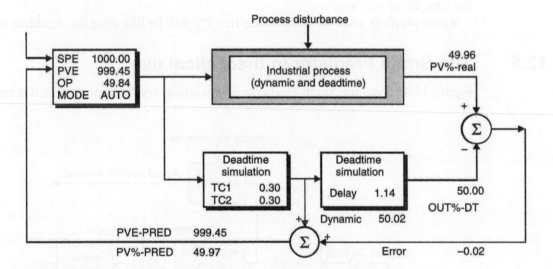

Figure 12.5
Block diagram of Exercise 14; closed loop control with process simulation

As there are no means of separating process deadtime from process dynamic in order to find out how the process would behave without deadtime, we make use of the values provided by a process simulation.

The process simulation is split into two parts as seen in Figure 12.5; these two parts are described in Sections 12.4.1 and 12.4.2.

13

Basic principles of fuzzy logic and neural networks

13.1   Objectives

This chapter serves to review the basic principles and descriptions of neural networks and fuzzy logic.

As a result of studying this chapter, the student should be able to:

- Describe the basic principles of fuzzy logic
- Describe the acronyms and basic terminology as used in neural networking and fuzzy logic applications.

13.2   Introduction to fuzzy logic

In the real world there is a lot of vague and imprecise conditions that defy a simple 'True' or 'False' statement as a description of their state. The computer and its binary logic are incapable of adequately representing these vague, yet understandable, states and conditions. Fuzzy logic is a branch of machine intelligence that helps computers understand the variations that occur in an uncertain and very vague world in which we exist.

Fuzzy logic 'manipulates' such vague concepts as 'warm' or 'going fast', in such a manner that it helps us to design things like air conditioners and speed control systems to move or switch from one set of control criteria to another, even when the reason to do so is because '*It is too warm*, or *not warm enough* to *go faster* or *slow down a bit*': all of these 'instructions' make sense to us, but are far removed from the digital world of just binary 1s and 0s.

The *true* and *false* statements that are absolute in there meaning come from a defined starting location and are designed to terminate at a known destination.

No known mathematical model can describe the action of, say a ship coming from some undefined point at sea into a dock area and finally coming to rest at a precise position on a wharf.

Humans and fuzzy logic can perform this action very accurately, if the wind blows a bit harder, or another ship hampers a particular docking maneuver this is sensed and an unrelated but effective action is taken. The action taken though is different each and every time as the disturbance is also different every time. (Similar but different events occur every time a ship tries to perform this procedure.)

When mathematicians lack specific algorithms that dictates how a system should respond to inputs, fuzzy logic can be used to either control or describe the system by using commonsense rules that refer to indefinite quantities.

Applications for fuzzy logic extend far beyond control systems; in principle they can extend to any continuous system, be it in, say physics or biology. It may well be that fuzzy models are more useful and accurate than standard mathematical ones.

13.3   What is fuzzy logic?

In standard set theory an object either *does* or *does not* belong to a set. There is '*no middle ground*'. This theory is an ancient Greek law propounded by Aristotle – the law of the excluded middle.

The number *five* belongs fully to the odd number set and not at all to the set of even numbers. In such bivalent sets an object cannot belong to both a set and its complement set or indeed to neither of the sets. This principle preserves the structure of logic and prevents an object being 'is' and 'is not' at the same time.

Sets that are 'fuzzy' or multivalent break this 'no middle ground' law to some extent. Items belong only partially to a fuzzy set, they may also belong to more than one set. The boundaries of standard sets are exact, those of fuzzy sets curve or taper off, and it is this fact that creates partial contradictions. The temperature of ambient air can be 20% cool and 80% not cool at the same time.

13.4   What does fuzzy logic do?

Fuzzy degrees are *not* the same as probability percentages. Probability measures whether something *will occur or not*. Fuzziness measures the *degree* to which *something occurs* or some *condition exists*.

13.5   The rules of fuzzy logic

The only real constraint in the use of fuzzy logic is that, for the object in question, its membership in complementary groups must sum to unity. If something is 30% cool, it must also be 70% *not* cool. This enables fuzzy logic to avoid the bivalent contradiction that something is 100% cool and 100% not cool, that would destroy formal logic (Figure 13.1).

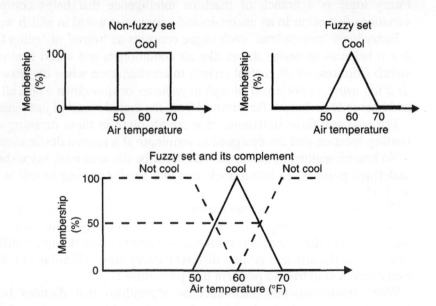

Figure 13.1
Fuzzy logic-complementary rules set

13.5.1   Fuzzy logic! a conundrum (Thanks to Bertrand Russell)

This section probably serves to illustrate the fuzzy logic world. Read this with your full attention though it illustrates the difference between half empty and half full!! It is a Greek paradox at the center of modern set theory and logic.

- A Cretan asserts that all Cretans lie.
- So, is he lying?
- If he lies, then he tells the truth and does not lie.
- If he does not lie then he tells the truth and so he lies.

Both cases lead to a contradiction because the statement is both true and false. The same paradox exists in set theory. The set of all sets is a set, so it is a member of itself. Yet the set of all apples is not a member of itself because its members are apples and not sets.
The underlying contradiction is then:

Is the set of all sets that are not members of themselves a member of itself?
If it is, it isn't; if it isn't, then it is.

Classic logic surrenders here, but fuzzy logic says that answer is half true and half false; a 50–50 divide 50% of the Cretans statements are true and 50% are false; he lies half the time and tells the truth for the other half. When membership is less than total a bivalent might simplify this by rounding it down to 0 or up to 100%. But 50% neither rounds up or down.

13.5.2   An example to illustrate fuzzy rules?

Fuzzy logic is based on the rules of the form 'If . . . Then' that converts inputs into outputs – one fuzzy set into another.
For example, the controller of a car's air conditioner might include rules such as:

- If the temperature is cool, then set the motor on slow.
- If the temperature is just right set the speed to medium.

The temperatures (cool and just right) and the motor speeds (slow and medium) name fuzzy sets rather than specific values.

13.5.3   Plotting and creating a fuzzy patch

If we now plot the inputs (temperature) along one axis of a graph, and the outputs (motor speed) along a second axis. The product of these fuzzy sets forms a fuzzy patch, an area that represents the set of all associations that the rule forms between those inputs and outputs. The size of this patch illustrates the magnitude of the rule's vagueness or uncertainty.
However, if 'cool' is precisely 21.5 °C the fuzzy set collapses to a 'spike'. If both the 'slow' and 'cool' sets are spikers, the rule patch is a point. This would be caused by 21.5 °C requiring a speed of 650 rpm for the motor – a logical result to this problem.

13.5.4   The use of fuzzy patches

A fuzzy system must have a set of overlapping patches that relate to the full range of inputs to outputs. It can be seen that enough small fuzzy patches can cover a graph of any function or input/output relationship. It is also possible to pick in advance the maximum error of the approximation and be sure there is a finite number of fuzzy rules that achieve it.

A fuzzy system 'reasons, or infers', based on its rule patches.

Two or more rules convert any incoming number into some result because the patches overlap. When data activates the rules, overlapping patches react in parallel – but only to some degree.

13.6   Fuzzy logic example using five rules and patches

As an example of fuzzy logic we will look at an example of an air conditioner that relies on five rules and therefore five patches to relate temperature to motor speed. Figure 13.2 illustrates this application.

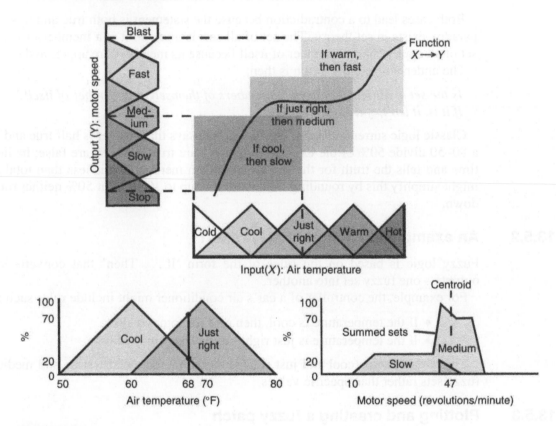

Figure 13.2
Application of fuzzy logic to the control of an air conditioner shows how manipulating vague sets can yield precise instructions

13.6.1   Defining the rules

The temperature sets are, *cold*, *cool*, *just right*, *warm* and *hot*; these cover all the possible fuzzy inputs.

The motor speed sets are *very slow*, *slow*, *medium*, *fast* and *maximum* describe all the fuzzy outputs.

A temperature of say 68 °F, might be represented by these fuzzy sets and rules:

- 20% cool and therefore 80% *not* cool and
- 70% just right and 30% *not* just right
- At the same time the air is 0% cold, warm and hot.

The 'if cool' and 'if just right' rules would fire and invoke both the *slow* and *medium* motor speeds.

13.6.2   Acting on the rules

The two rules contribute proportionally to the final motor speed. Because the temperature was 20% cool, the curve describing the slow motor must shrink to 20% of its height. The medium curve must shrink to 70% for the same reason. Summing these two curves results in the final curve for the fuzzy set.

However, in its fuzzy form it cannot be understood by a binary system so the final step in the process is defuzzification.

13.6.3   Defuzzification

This is where the resultant fuzzy output curve is converted to a single numeric value. The normal way of achieving this is by computing the center of mass, or centroid, of the area under the curve. In our example, this corresponds to a speed of 47 rpm.

Thus, beginning with a quantitative temperature input, the electronic controller can reason from fuzzy temperature and motor speed sets and arrives at an appropriate and precise speed output.

13.7   The Achilles heel of fuzzy logic

The weakness of fuzzy logic is its rules. These are, in the majority of fuzzy applications, set by engineers who are expert in the related application. This leads to a lengthy process of 'tuning' these rules and the fuzzy sets. To automate this process many companies are resorting to building and using adaptive fuzzy systems that use neural networks or other statistical tools to refine or even form those initial rules.

13.8   Neural networks

Neural networks are collections of 'neurons' and 'synapses' that change their values in response from inputs from surrounding neurons and synapses. The neural net acts like a computer because it maps inputs to outputs. The neurons and synapses may be silicon components or software equations that simulate their behavior. A neuron sums all incoming signals from other neurons and then emits its own response in the form of a number. Signals travel across the synapses that have numerical values that weigh the flow of neuronic values.

When new input data 'fires' a network of neurons, the synaptic values can change slightly. A neural net 'learns' when it changes the value of its synapsis.

13.8.1   The learning process

Figure 13.3 illustrates a typical neural network and its synaptic connections. Each neuron, as illustrated, receives a number of inputs X_I which are assigned weights, by the interconnecting synapse W_I. From the weighted total input, the processing element computes a single output value Y.

Figure 13.4 shows what occurs inside each neuron when it is activated or processed.

Neuron inputs

Various signals (inputs of the neuron X_I) are received from other neurons via synapses.

Neuron input calculation

A weighted sum of these input values is calculated.

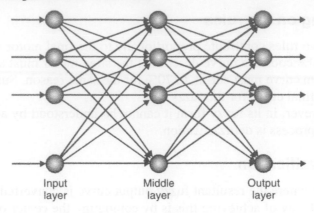

Figure 13.3
Typical neural network connections

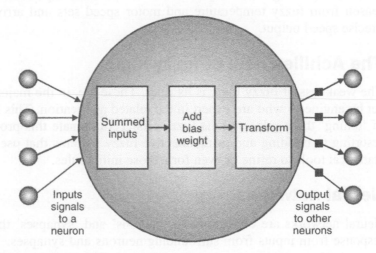

Figure 13.4
Processing steps inside a neuron

Neuron internal function transform

The sum of the input calculation is transformed by an internal function, which is normally, but not always, fixed for a neuron at the time the network is constructed.

Neuron output

The transformed result (outputs) are sent individually onto other neurons via the interconnecting synapse.

13.8.2   Neuron actions in 'learning'

Learning implies that the neuron changes its input to output behavior in response to the environment in which it exists. However, the neurons transform function is usually fixed, so the only way the input-to-output transform can be changed is by changing the bias weight as applied to the *inputs*.

So 'learning' is achieved by changing the weights on the inputs, and the internal model of the network is embodied in the set of all these weights. How are these 'weights' changed? One of the most common forms widely used is called 'back propagation networking', commonly used for chemical engineering.

13.9   Neural back propagation networking

These networks always consist of three neuron layers: input, middle and output layer. The construction is such that a neuron in each layer is connected to every neuron in the next layer (Figure 13.3). The number of middle layer neurons varies, but has to be selected with care; too many result in unmanageable patterns, and too few will require an excessive number of iterations to take place before an acceptable output is obtained.

13.9.1   Forward output flow (neuron initialization)

The initial pattern of neuron weights is randomized and presented to the input layer, which in turn passes it on to the middle layer. Each neuron computes its output signal as

$$I = WX + B$$

The $\sum$ output Ij is determined by multiplying each input signal by the random weight value on the synoptical interconnection:

$$Ij = \sum_i W_{ij-1} X_{ij-1} + B_j$$

This weighted sum is transformed by a function, $f(X)$ called the activated function of the neuron and it determines the activity generated in the neuron as a result of an input signal of a particular size.

13.9.2   Neural sigmoidal functions

For back propagation networks, and for most chemical engineering applications, the function described in Section 13.9 is a sigmoidal function. This function, as shown in Figure 13.5, is:

- Continuous
- S shaped
- Monotonically increasing
- Asymptotically approaches fixed values as the input approaches $\pm \infty$.

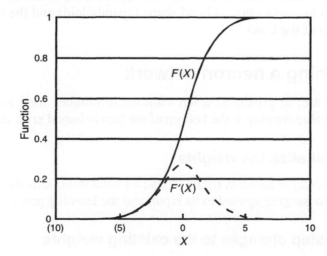

Figure 13.5
A sigmoidal function

Generally, the upper limit of a sigmoid is set to +1 and the lower limit to 0 or −1. The steepness of the curve and even the exact function used to compute it is generally less important than the general 'S' shape.

The following sigmoidal curve expressed as a function of I_j, the weighted input to the neuron is widely used.

$$X_j = f\left(I_j\right) = \frac{1}{1 + e^{-(I_j + T)}}$$

Where T is a simple threshold and X is the input. This transformed input signal becomes the total activation of the middle layer neurons, which is used for their outputs, and which in turn become the inputs to the output neuron layer, where a similar action takes place in the output neuron, using the sigmoidal function which produces a final output value from the neuron.

13.9.3   Backward error propagation (the delta rule)

The result of the output is compared with the desired output. The difference (or error) becomes a bias value by which we modify the weights in the neuron connections. It usually takes several iterations to match the target output required.

The delta method is one of the most common methods used in backward propagation. The delta rule iteratively minimizes the average squared error between the outputs of each neuron and the target value. The error gradient is then determined for the hidden (middle) layers by calculating the weighted error of that layer. Thus,

- The errors are propagated back one layer
- The same procedure is applied recursively until the input layer is reached
- This is backward error flow.

The calculated error gradients are then used to update the network weights. A momentum term can be introduced into this procedure to determine the effect of previous weight changes on present weight changes in weight space. This usually helps to improve convergence.

Thus back propagation is a gradient descent algorithm that tries to minimize the average squared error of the network by moving down the gradient of the error curve. In a simple system the error curve is bowl shaped (paraboloid) and the network eventually gets to the bottom of the bowl.

13.10   Training a neuron network

There are, in principle, seven standard techniques used to 'train' a network for a zero error value (resting at the bottom of the bowl-shaped error curve).

13.10.1   Re-initialize the weights

If difficulty is found in trying to find a global minimum, the process can have a new set of random weights applied to its input, and the learning process repeated.

13.10.2   Add step changes to the existing weights

It is possible for the network to 'oscillate' around an error value due to the fact that any calculated weight change in the network does not improve (decrease) the error term.

All that is normally needed is for a 'slight push', to be given to the weighted factors. This can be achieved by randomly moving the weight values to new positions, but not too far from the point of origin.

13.10.3   Avoiding overparameterization

If there are too many neurons in the middle, or hidden, layer then overparameterization occurs which in turn gives poor predictions. Reduction of neurons in this layer affects a cure to this problem. There is no rule of thumb here, and the number of neurons needed is best determined experimentally.

13.10.4   Changing the momentum term

This is the easiest thing to do if the system is a software network, the momentum term α is implemented by adding a part of the last weighted term to the new one, changing this value, again best done experimentally, can assist with a cure.

13.10.5   Noise and repeated data

Avoid repeated or less noisy data

Repeated or noise-free inputs makes the network remember the pattern rather than generalizing their features. If the network never sees the same input values twice, this prevents it from remembering the pattern. Introducing noise can assist in preventing this from occurring.

13.10.6   Changing the learning tolerance

The training of a network ceases once the error value for all cases is equal or less than the learning tolerance. If this tolerance is too small, the learning process never ceases. Experiment with the tolerance level until a satisfactory point is reached where the weights cease changing their value in any significant way.

13.10.7   Increasing the middle (hidden) layer value

This is the inverse of the problem described in Section 13.10.3, 'avoiding overparameterization' and is used if-all-else-fails. In other words we have too few neurons in the middle layer, where as in Section 13.10.3 we had too many. In general an increase >10% shows improvements.

13.11   Conclusions and then the next step

Fuzzy logic has been used since the early 1980s and has been very successful in many applications, such as Hitachi's use of it in controlling subway trains, and it proving so accurate and reliable that its performance exceeds what a trained (no pun intended) driver can do. When its principles are used with neuron, or self-learning networks, we have a very formidable set of tools being made available to us. When this is applied to process control systems, it gives us a foreseeable future of control systems that are error free, and can cope with all the variances, including the operator, to such an extent that we could finally have a 'perfect' system.

Meanwhile, we have two more issues to look at that may well be considered as the intermediate step from the past and current technology to the ultimate, self-learning and totally accurate control system. These two issues, discussed and described in Chapter 14, conclude this book on practical process control. They are statistical process control (SPC) and self-tuning controllers. SPC is used to see where the process is in error, a problem solved by neuron networks, and self-tuning controllers, a problem that can be solved by fuzzy logic.

14

Self-tuning intelligent control and statistical process control

14.1 Objectives

As a result of studying this chapter, the student should be able to:

- Describe the theory and operation of a self-tuning controller
- Describe the concept of statistical process control (SPC) and its use in analyzing and indicating the standards of performance in control systems.

This chapter introduces the basic concepts of self-tuning or adaptive controllers, intelligent controllers, and provides an overview of statistical process control (SPC).

14.2 Self-tuning controllers

Self or auto-tuning controllers are capable of automatically re-adjusting the process controllers tuning parameters. They first appeared on the market in the early 1970s and evolved from ones using optimum regulating and control types through to the current types that, with the advent of high speed processors, rely on adaptive control algorithms.

The main elements of a self-tuning system are illustrated in Figure 14.1, these being:

- *A system identifier*:   This model estimates the parameters of the process.
- *A controller synthesizer*:   This model has to synthesize or calculate the controller parameters specified by the control object functions.
- *A controller implementation block*:   This is the controller whose parameters (gain K_C, T_{INT}, T_{DER}, etc.) are changed and modified at periodic intervals by the controller synthesizer.

14.2.1 The system identifier

The system identifier, by comparing the PV action as a result of the MV change, and using algorithms based on recursive estimations, determines the response of the system. This is commonly achieved by the use of fuzzy logic that extracts key dynamic response features from the transient excursion in the system dynamics. These excursions may be deliberately invoked by the controller, but are usually the start-up, and ones caused by process disturbances that are the normal ones.

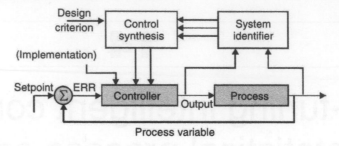

Figure 14.1
The main components of a self-tuning system

14.2.2   The controller synthesizer

The desired values for the PI and D algorithms used by the controller are determined by this block. The calculations can vary from simple to highly complex, depending on the rules used.

14.2.3   Self-tuning operation

This technique requires a starting point derived from knowledge of known and operational plants of a similar nature. This method is affected by the relationship between plant and controller parameters. Since the plant parameters are unknown they are obtained by the use of recursive parameter identification algorithms. The control parameters are then obtained from estimation of the plant parameters.

Referring to Figure 14.1, the controller is called 'self-tuning' since it has the ability to tune its own parameters. It consists of two loops, an inner one that is the conventional control loop having however varying parameters, and an outer one consisting of an identifier and control synthesizer which adjust the process controllers parameters.

14.3   Gain scheduling controller

Gain scheduling control relies on the fact that auxiliary or alternate process variables are found that correlate well with the main process variable. By taking these alternate process variables it is possible to compensate for process variations by changing the parameter settings of the controller as functions within the auxiliary variables. Figure 14.2 illustrates this concept.

14.3.1   Gain scheduling advantages

The main advantage is that the parameters can be changed quite quickly in response to changes in plant dynamics. It is convenient if the plant dynamics are simple and well-known.

14.3.2   Gain scheduling disadvantages

The disadvantage is that the gain scheduling is an *open loop* adaptation and has no real learning or intelligence. The extent and design criteria can be very large too. Selection of the auxiliary point of measurement has to be done with a great deal of knowledge and thought regarding the process operation.

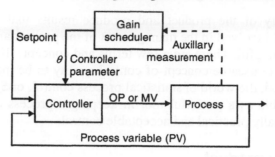

Figure 14.2
Gain scheduling controller

14.4 Implementation requirements for self-tuning controllers

Self-tuning controllers that deliberately introduce known disturbances into a system in order to measure the effect from a known cause are not popular. Preference is given to self-tuning controllers that sit in the background, measure and evaluate what the control controller is doing. Then comparing this with the effect this has on the process, and making decisions on these measured parameters, the controllers-operating variables are updated.

To achieve this, the updating algorithms are usually kept dormant until the error term generated by the system controller becomes unacceptably high (>1%) at which point the correcting algorithms can be unleashed on the control system, and corrective action can commence.

After the error has evolved, the self-tuning algorithm can check the response of the controller in terms of the period of oscillation, damping and overshoot values. Figure 14.3 illustrates these parameters.

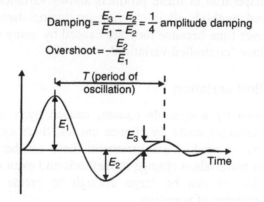

$$\text{Damping} = \frac{E_3 - E_2}{E_1 - E_2} = \frac{1}{4} \text{ amplitude damping}$$

$$\text{Overshoot} = -\frac{E_2}{E_1}$$

Figure 14.3
Controller measurement parameters

14.5 Statistical process control (SPC)

The ultimate objective of a process control system is to keep the product, the final result as produced by the process, always within all the pre-defined limits set by the products description.

There are almost an infinite number of methods and systematic approaches available in the real engineering world to help achieve this. However, although all these tools exist, it is necessary to have procedures that analyze the process's performance, compare this with the quality of the product and produce results that are both 'understandable' by all personnel involved in the management of the process and of course are also both accurate and meaningful. There are a few terms and concepts that need to be understood to enable a basic and useable concept of control quality to be managed, and once these have been understood, the world of statistical process control, or SPC becomes apparent, meaningful and useable as a powerful tool in keeping a process control system under economical, operationally practical and acceptable control.

14.5.1   Uniform product

Only by understanding the process with all of its variations and quirks, product disturbances, hiccups and getting to know its individual 'personality' can we hope to achieve a state of virtually uniform product. No two 'identical' plants or systems will ever produce identical product, similar, yes, but never identical. This is where SPC helps in identifying the 'identical' differences.

Dr Shewhart, working at the Bell Laboratories in the early 1920s, after comparisons made between variations in nature and items produced in process systems found inconsistencies and formulated the following statement:

While every process displays variation,

- Some processes display *controlled variation*
- Others display *uncontrolled variation*.

Controlled variation

This is characterized by a stable and consistent pattern of variation over time, attributable to 'chance' causes. Consider a product with a measurable dimension or characteristic (mechanical or chemical). Samples are taken in the course of a production run. The results of inspection of these products shows variances caused by machines, materials, operators and methods all interacting producing these variations. These variations are consistent over time because they are caused by many contributing factors. These chance causes produce 'controlled variation'.

Uncontrolled variation

This is caused by assignable causes, caused by a pattern of variation over time. In addition to changes made by chance causes, there exists special factors that can cause large impacts on product measurement; these can be caused by maladjusted machines, difference in materials, a change in methods end even changes in the environment. These assignable factors can be large enough to create marked changes in known and understood patterns of variation.

14.6   Two ways to improve a production process

The two methods described here to improve a process are fundamentally different, one looks for change to a consistent process, the other for modifications to the process.

14.6.1   Controlled variations problem

When a process displays controlled variation it should be considered stable and consistent. The variations are caused by factors inherent in the actual process. To reduce these variations it will be necessary to change the process.

14.6.2 Uncontrolled variations problem

This means the process is varying from time to time. It is both inconsistent and unstable. The solution is to identify and remove the cause(s) of the problem(s).

14.7 Obtaining the information required for SPC

There is fundamentally only one way to record the real processes performance, and that is by a *strip chart* showing the process variable signal, and possibly the controller output signals. That is, the commands sent into the process and the processes reply to these commands in both *magnitude and time*.

14.7.1 Statistical inference

The average of 2, 4, 6 and 8 is 5 this being the balance point for this sample of data values. The sample range for this data is 6, that is how far apart 2 and 8 are (the maxima and minima). However statistical inference relies on the fact that a conceptual population exists, this being needed to enable us to rationalize any attempt at prediction and that all samples taken were from this population, this being needed to believe in the estimates based on the sample statistics.

For the sake of simplicity and clarity we will consider that all samples are objective and represent one conceptual population.

If this is not true then the results may well be inconsistent and the statistics will be erratic. In fact, if this happens, the process can be considered schizophrenic; thus the process is displaying uncontrolled variation. The resultant statistics simply could not be generalized.

14.7.2 Using sub-groups to monitor the process

Each sample collected at a single point in time is a sub-group, each one being treated as a separate sample. Figure 14.4 shows four sub-groups selected from a stable process, one sub-group per hour. The bell-shaped profiles represent the total output of the process each hour, the dots representing the measurements taken in each group.

8 9 10 11

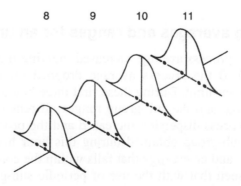

Figure 14.4
Four sub-groups selected from a stable system

14.7.3   Recording averages and ranges for a stable process

The next step is to record the average and ranges onto a time-scaled strip chart, this being shown in Figure 14.5. As long as these plots move around within the defined upper and lower limits displayed also on the chart, we can consider that all sub-groups were derived from the same conceptual population.

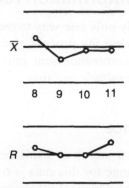

Figure 14.5
The average *and* range *chart for a stable process*

If we now consider the same example, but the process itself is changing from hour to hour, i.e. there is variation in the process. We let the bell-shaped curves in the Figure 14.6 represent the variation in the processes output each hour.

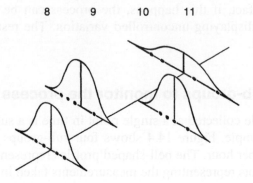

Figure 14.6
Four sub-groups from an unstable *system*

14.7.4   Recording averages and ranges for an unstable process

At 09:00 the process average increased, moving the sub-group average above the upper limit. At 10:00 the process average dropped dramatically, and the sub-group moved below the lower limit. During these first three hours, 08:00–11:00, the process dispersion did *not* change and the sub-group ranges all remained within the control limits. But at 11:00, the process dispersion increased and the process average moved back to its initial value. The sub-group obtained during this hour has a *range* that falls above the upper control limit, and an *average* that falls within the control limits.

It can be seen that with the use of periodic sub-groups, two additional variables have been introduced, namely – the sub-group average and the sub-group range. *These are the two variables used to monitor the process.*

The following example illustrates the behavior of these two variables and how they relate to the measurements when the process is stable. Refer to Table 14.1 and Figure 14.7.

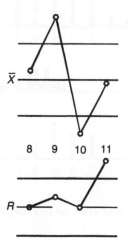

Figure 14.7
Illustrating the average *and* range *chart for an* unstable *process*

The data in this example is taken from a stable process. The measurements shown represent the thickness of a product. The numbers represent how much the part exceeded 0.300 in., in 0.001 units.

Sub-group Number	Variance	Average	Range	Sub-group Number	Variance	Average	Range
1	1 4 6 4	3.75	5	11	4 5 6 5	5.00	2
2	3 7 5 5	5.00	4	12	6 7 8 5	6.50	3
3	4 5 5 7	5.25	3	13	3 3 7 3	4.00	4
4	6 2 4 5	4.25	4	14	6 3 2 9	5.00	7
5	1 6 7 3	4.25	6	15	7 3 4 3	4.25	4
6	8 3 6 4	5.25	5	16	6 4 6 5	5.25	2
7	7 5 6 6	6.00	2	17	5 5 0 5	3.75	5
8	5 3 4 6	4.50	3	18	6 4 6 3	4.75	3
9	4 5 9 2	5.00	7	19	6 4 4 0	3.50	6
10	7 5 6 5	5.75	2	20	6 2 5 4	4.25	4

Table 14.1
Data for Figure 14.7

14.7.5   Example of a stable process

The histograms in Figure 14.8 are all to scale on both axes. However all three represent totally different profiles and dispersions. It is therefore essential to distinguish between these variables.

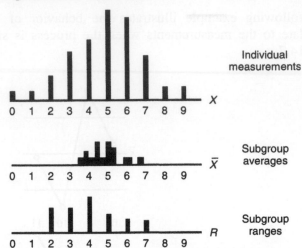

Figure 14.8
Histograms and measurements, averages and ranges for Table 14.1

14.7.6   Distributions of measurements, averages and ranges

While the measurements, averages and ranges have different distributions, these are related in certain ways when they are derived from a stable process. Figure 14.9 shows these relationships more clearly.

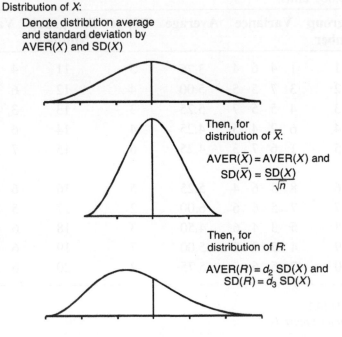

Figure 14.9
Distributions of measurements, averages and ranges

Notations related to Figure 14.9:

Let AVER(X) denote the average of the distribution of the X values. Let SD(X) denote the standard deviation of the distribution of X.

In a similar manner AVER($\overline{X}$) and SD($\overline{X}$) denote the average and standard deviation of the distribution of the sub-group averages, while AVER(R) and SD(R) denote the ranges.

With this notation the relationships between the averages and standard deviations can be expressed as:

$$\text{AVER}(\overline{X}) = \text{AVER}(X)$$

$$\text{SD}(\overline{X}) = \frac{\text{SD}(X)}{\sqrt{n}}$$

$$\text{AVER}(R) = d_2 \times \text{SD}(X)$$

$$\text{SD}(R) = d_3 \times \text{SD}(X)$$

Constants d_2 and d_3 are scaling factors that depend on the sub-group size n. These factors are shown in Table 14.2 and are based on the normality of X.

n	d_2	d_3	n	d_2	d_3
2	1.128	0.853	14	3.407	0.762
3	1.693	0.888	15	3.472	0.755
4	2.059	0.880	16	3.532	0.749
5	2.326	0.864	17	3.588	0.743
6	2.534	0.848	18	3.640	0.738
7	2.704	0.833	19	3.689	0.733
8	2.847	0.820	20	3.735	0.729
9	2.970	0.808	21	3.778	0.724
10	3.078	0.797	22	3.819	0.720
11	3.173	0.787	23	3.858	0.716
12	3.258	0.778	24	3.895	0.712
13	3.336	0.770	25	3.931	0.709

Table 14.2
Factors for the average and standard range deviation of the range distribution

14.8 Calculating control limits

From the four relationships that have been shown above it is possible to obtain control limits for the sub-group averages and ranges.

There are two principle methods of calculating control limits, one is the structural approach and the other is by formulae. Both methods are illustrated in Sections 14.8.1 and 14.8.2.

When first obtaining control limits it is customary to collect 20–30 sub-groups before calculating the limits. By using many sub-groups the impact of an extreme value is minimized.

Using the example illustrated in Section 14.8.5 of sub-group averages and range values from a control process, the next two sections serve to illustrate the structural and formulated approach in calculating the process control limits.

Using the data shown in Table 14.3, the control limits, as shown in Figure 14.9 will be found using both structural and formulated approaches.

Sub-Group Size	A_2	D_3	D_4
2	1.880	0.000	3.268
3	1.023	0.000	2.574
4	0.729	0.000	2.282
5	0.577	0.000	2.114
6	0.483	0.000	2.004
7	0.419	0.076	1.924
8	0.373	0.136	1.864
9	0.337	0.184	1.816
10	0.308	0.223	1.777

Table 14.3
Calculating control limits for example 14.8.5

14.8.1   The structural approach

1. First we estimate the distribution of the X values:

$$\text{The grand average } \overline{\overline{X}} = 4.763 \text{ estimates AVER}(X)$$
$$\text{The average range is } \overline{R} = 4.05$$
$$\text{So the estimated of SD}(X) \text{ is } \frac{\overline{R}}{d_2} = 1.967$$

2. Estimation of the distribution of the $\overline{X}$ values:

$$\text{The grand average } \overline{\overline{X}} = 4.763 \text{ estimates AVER}(\overline{X})$$
$$\text{The estimate of SD}(\overline{X}) \text{ is } \frac{\dfrac{\overline{R}}{d_2}}{\sqrt{n}} = \frac{1.967}{2} = 0.984$$

3. Control for sub-group averages are:

$$4.763 \pm 3(0.984) = 1.811 - 7.715$$

4. Estimates for the distribution of R values:

$$\text{The average range } \overline{R} = 4.05 \text{ estimates AVER}(R)$$
$$\text{The estimate of SD}(R) \text{ is } \frac{d_3 \times \overline{R}}{d_2} = \frac{0.88 \times 4.05}{2.059} = 1.731$$

5. Control limits for sub-group ranges are:

$$4.05 \pm 3(1.731) = -1.143 - 9.423$$

Since the sub-group ranges are non-negative, the negative lower limit has no meaning. In this case the lower control limit = 0.

14.8.2   The formulated approach

- The grand average is 4.763
- The average range is 4.05
- The sub-group size is 4.

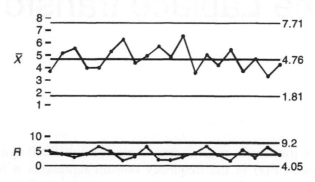

Figure 14.10
The resultant control chart for the worked example in Section 14.8.5

14.9   The logic behind control charts

In conclusion, both to this chapter and to the workshop, the following Figure 14.11 illustrates the logic behind control charts.

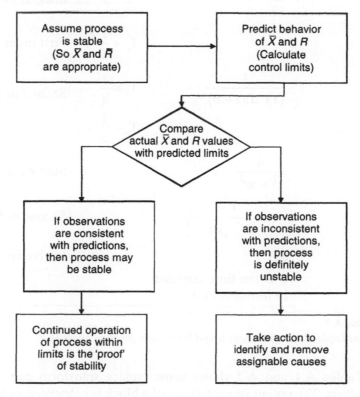

Figure 14.11
The logic behind control charts

Appendix A

Some Laplace transform pairs

Laplace transforms make it easy to represent difficult dynamic systems. A mathematical expression $F(s)$ in the frequency domain represents a function in the time domain, a transfer function $F(t)$ or a time function $f(t)$. A transfer function represents the properties (or the behavior) of a mathematical block (or calculation). A time function represents a value (or signal) over time.

$F(s)$		Block type
1		Gain block (gain = 1)
$\dfrac{1}{sT}$		Integral block
$\dfrac{1}{s+a}$	$T_a = \dfrac{1}{a}$	First order lag
$\dfrac{1}{(s+a)(s+b)}$	$T_a = \dfrac{1}{a}$ $T_b = \dfrac{1}{b}$	Second order lag
$\dfrac{w}{s^2 + w^2}$		Sine wave (2 integrators)
$\dfrac{1}{s^2 + 2\,{}_n s + {}_n^2}$		Second order system
ST		Derivative block
T is the time constant in formulas F(s)		

Table A.1
Some Laplace transform pairs useful for transfer function analysis

Tables A.1 and A.2 shows some laplace transform pairs useful for control system analysis. The output signal $f(s)_{output}$ of a block is calculated as follows:

$$f(s)_{output} = F(s)_{block} \times f(s)_{input}$$

An explanation of laplace transform theorems is beyond the scope of this publication and not intended.[1] Two examples will be given in Figures A.1 and A.2.

$F(s)$	$f(t)$
1	Unit impulse
$\dfrac{1}{s}$	Unit step
$\dfrac{1}{s^2}$	Unit ramp
$\dfrac{1}{s+a}$	e^{-at}
$\dfrac{1}{s(s+a)}$	$\dfrac{1}{a}(1-e^{-at})$
$\dfrac{1}{(s+a)(s+b)}$	$\dfrac{1}{b-a}(e^{-at}-e^{-bt})$
$\dfrac{w}{s^2+w^2}$	$\sin wt$
$\dfrac{1}{s^2+2dw_n s + w_n^2}$	$\dfrac{1}{w_d}e^{-dw_n t}\sin w_d t$ $w_d \stackrel{\circ}{=} w_n \sqrt{1-d^2}$
$\dfrac{1}{s(s^2+2dw_n s + w_n^2)}$	$\dfrac{1}{w_d^2}-\dfrac{1}{w_n w_d}e^{-dw_n t}\sin(w_d t + F)$ $w_d \stackrel{\circ}{=} w_n \sqrt{1-d^2}$ $F \stackrel{\circ}{=} \cos^{-1} d$

Table A.2
Some Laplace transform pairs useful for time function analysis f (t)

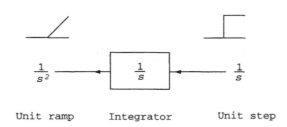

Unit ramp   Integrator   Unit step

Figure A.1
Integral block with step input

---

1. For further studies read 'Feedback and Control Systems' by DiStefano III, Stubberud and Williams, published by McGraw-Hill Book Company as part of Schaum's Outline Series.

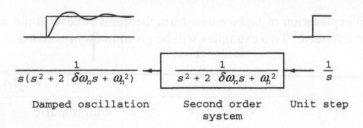

Damped oscillation   Second order   Unit step
                              system

Figure A.2
Second order system with step input

The integral block and its input, a step function, is a good example to show that the same function $1/s$ in the frequency domain may represent an integral calculation (block or transfer function) or a step function (input signal).

A second order system is a close representation of the behavior of many industrial processes

Appendix B

Block diagram transformation theorems

Complicated block diagrams can be broken into several easily recognized blocks. The summary of transformation theorems is a useful tool for this. W, X, Y, Z represent signals $f(s)$ in the frequency domain. P, P_1 and P_2 represent transfer functions $F(s)$.

$$Y = (P_1 P_2) X$$

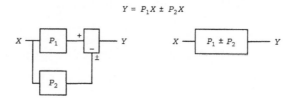

Figure B.1
Blocks in cascade

$$Y = P_1 X \pm P_2 X$$

Figure B.2
Parallel blocks

$$Y = \frac{P_1}{P_1 X \pm P_2} X$$

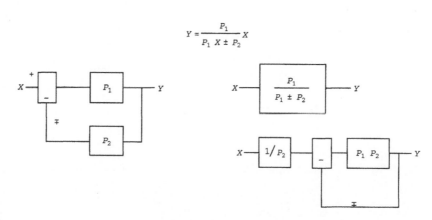

Figure B.3
Feedback loop

$$Z = P(X \pm Y)$$

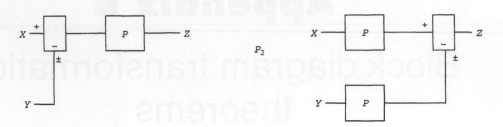

Figure B.4
Moving summing point

$$Y = PX$$

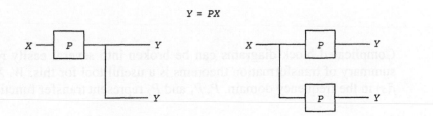

Figure B.5
Moving take-off point

Appendix C

Detail display

Each detail display is designed as operator interface for one controller or one major control unit. It displays and permits operation on variables, parameters and limits. The detail displays in the training applications use a similar layout and representation of information as with many industrial operator workstations (Figure C.1).

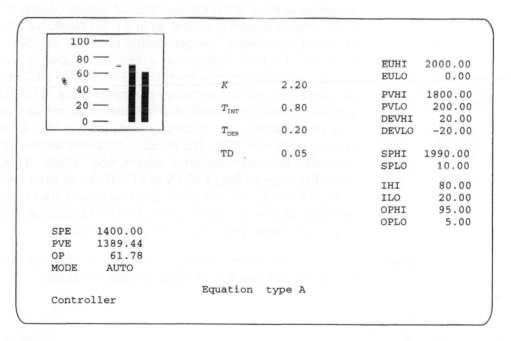

Figure C.1
Detail display

The detail display is divided vertically into three parts, a left, middle and right section. The section on the left hand side is for *major control* variables and bar-graphs. The section in the middle is for *tuning constants*. The section on the right hand side is for *range*, *limit*, *alarm* and *status parameters*.

An explanation of the abbreviations (or acronyms) used in the manual and on the displays are explained in the sections below.

C.1   **Major control variables**

SPE   The setpoint in engineering units or the target value for the controller. The objective is to control and keep the value of PVE at the value of SPE. 0% of range of the SPE is defined by EULO and 100% of range of SPE is defined by EUHI. A green marker, on the left side of bar-graphs represents the SPE in % of range.

PVE   Process variable in engineering units. The controller is to control and keep the PVE at the value of SPE. 0% of range of PVE is defined by EULO and 100% of range of PVE is defined by EUHI. A cyan bar-graph represents the PVE in % of range.

OP   Output of controller in % to manipulate the process or the setpoint of a secondary controller. A yellow bar-graph represents the OP in % of range. The human operator can change the value of OP in MANUAL mode only.

MODE   Status variable to define the MODE of operation of a controller. MODE can assume the status MANUAL, AUTO, CASC, I-MAN, I-AUTO or I-CASC. MANUAL mode disables automatic control of the OP and enables the operator to manipulate the OP manually. AUTO mode enables automatic control of the OP and manual setting of SPE. CASC (cascade) mode enables automatic control, as in AUTO mode, but no operator setting of SPE is possible if another controller drives the SPE as its primary controller. If two conditions in a cascade control system are true, the primary controller assumes the MODE I-MAN, I-AUTO or I-CASC. These two conditions are: the secondary controller is not in CASC mode and the secondary controller is configured to initialize the primary controller.

In this situation, the secondary controller's SPE is used to set the primary controller's OP. The primary controller waits until the secondary controller has been switched into CASC mode. Then, the primary controller changes from I-MAN to MANUAL or from I-AUTO to AUTO or from I-CASC to CASC. The mode indicated after I, as in I-AUTO, is called pending mode. For example, I-AUTO means the primary controller is presently initialized with AUTO mode pending.

Tag name   A unique controller identification name. In the left bottom corner of Figure C.1, we find the generic descriptor 'controller'.

C.2   **Tuning constants**

K Controller gain. Generally, K is an overall controller gain for proportional, integral and derivative control.

                Special case:
                For $K = 0$, proportional and derivative control are disabled. Integral control works with unit gain of 1.

T_{INT} Time constant for integral control (in minutes). The following example will give some idea of the meaning of T_{INT}: If the input of the integrator is a step of magnitude Y, the OP is a ramp function with a slope of $K \times Y/T_{INT}$.

T_{DER} Time constant for derivative control (in minutes). The control is based on 'rate of change of input' multiplied by K.

TD   Time constant for digital filtering of PVE. This is the time constant of a low-pass filter between the field input of a process variable (raw value of PV) and the PVE. The purpose is to filter out unwanted noise which has no bearing on the true behavior of PVE.

Note: Every field controller has an analog input filter as well. This analog filter has the purpose of filtering out all frequencies in an input signal that are too high for calculations within the controller's scan time. If these high frequencies have to be processed, a special controller with a very short scan time has to be selected.

This problem is referred to as aliasing (and states that the minimum sampling frequency must be at least twice the higher frequency component of the PV, or errors will occur in the representation of the digital data).

C.3   Range, limit, alarm and status parameters

EUHI and EULO   High and low limits of the operational range of SPE and PVE in engineering units. Generally, a field input is transmitted to the controller as a 4–20 mA (milli-ampere) signal. The controller has to know the range of measurement. For example, 4 mA (0% of range) may represent 0 C and 20 mA (100% of range) may represent 2000 C.

PVHI and PVLO   Alarm limits in engineering units, setting and displaying an alarm status within the controller. A PVHI alarm is set, if PVE is above PVHI. A PVLO alarm is set, if PVE is below PVLO. The value of PVE itself is not limited.

DEVHI and DEVLO   Deviation alarm limits in % of range, setting and displaying an alarm status within the controller when PVE deviates by too much from the SPE. A DEVHI alarm is set, if (PVE – SPE) is above DEVHI. A DEVLO alarm is set, if (PVE – SPE) is below DEVLO. The value of PVE itself is not limited.

SPHI and SPLO   Setpoint high or setpoint low. This clamps the value of *SPE* at the limits SPHI and SPLO, when a primary controller attempts to drive SPE beyond SPHI or SPLO. Generally, industrial controllers permit the human operator to set *SPE* within full range, independently of SPHI and SPLO. SPHI and SPLO are in engineering units.

IHI and ILO   Output limits for integral control only (in %). An IHI alarm is set and the integral calculation is suspended when both of the following are true: the value of OP is above IHI, and further integration would have increased OP (integral wind-up high). An ILO alarm is set and integral calculation is suspended when both of the following are true: the value of OP is below ILO, and further integration would have decreased OP (integral wind-up low). Proportional and derivative control is not affected at all by IHI or ILO.

OPHI and OPLO	Output high and output low. OPHI and OPLO are in % and clamp the value of the OP. OPHI and OPLO have priority over IHI and ILO. Generally, industrial controllers permit the human operator to set OP within full range, independent of OPHI and OPLO.
CONFIG	Status variable, defining initialization and/or PV-tracking. INIT initializes an upstream primary controller's OP if the initializing controller (secondary) is not in CASC mode. *Tracking* forces SPE to assume (track) the value of PVE if the controller is not in AUTO or CASC mode. I and TR perform both initialization and PV-tracking.
EQUATION	Status variable, defining different selections of inputs for PID control: Type A calculates PID-control on ERR (ERR = PVE – SPE). Type B calculates PI-control on ERR (ERR = PVE – SPE) and D-control on PVE only. Type C calculates I-control on ERR (ERR = PVE – SPE) and PD-control on PVE only.

Appendix D

Auxiliary display

Display page (F8) holds an auxiliary display with all variables necessary for process simulation and control. It also shows those control variables, normally not shown on real industrial control displays, that are helpful in the understanding of process control. This display shows variables such as simulation gain and values, noise gain and values, disturbance gain and values, special controller output and limit calculations, etc.

D.1   Controller variables

OP	Controller output in % of range.
PV	Process variable in % of range.
SP	Setpoint in % of range.
CSP	Internal computed setpoint in % of range.
CVPD	CVPD is the increment value that occurs as a result of proportional and derivative control calculation. CVPD is used to increment (or decrement) all controller outputs by its value.
CVI	CVI is the increment value that occurs as a result of integral control calculation. CVI is used to increment (or decrement) all controller outputs by its value.
OPCALC	Status variable to define the type of OP-limit calculation. OPCALC can be set to the status VIRTUAL or REAL. VIRTUAL permits the internal output value (OPVIRT) to assume values beyond any OP-limit. The real output OP is then limited by OP-limits.

Any internal controller calculation makes use of the virtual, internal and unlimited output value. This results in a saturated OP limit calculation. REAL causes both to be limited by OP-limits, the internal output value (OPVIRT) and the true OP. This results in a *non*-saturated OP limit calculation. |
| ACTION | Direction of output control action. ACTION can be set to REVERSE or DIRECT. With DIRECT control action, the output value moves in |

the same direction as the PV value. With REVERSE control action, the output value moves in opposite direction to the PV value.

INIT
This status variable shows whether a controller is initialized by another downstream controller.

MODE
Status variable to define the MODE of operation of a controller. MODE can assume the status MANUAL, AUTO, CASC, I-MAN, I-AUTO or I-CASC. MANUAL mode disables automatic control of OP and enables the operator to manipulate OP manually. AUTO mode enables automatic control of OP and manual setting of SPE. CASC (cascade) mode enables automatic control, as in AUTO mode, but no operator setting of SPE is possible if another controller drives SPE as its primary controller. If two conditions in a cascade control system are true, the primary controller assumes the MODE I-MAN, I-AUTO or I-CASC. These two conditions are: the secondary controller is not in CASC mode and the secondary controller is configured to initialize the primary controller.

In this situation, the secondary controller's SPE is used to set the primary controller's OP. The primary controller waits until the secondary controller has been switched into CASC mode. Then, the primary controller changes from I-MAN to MANUAL or from I-AUTO to AUTO or from I-CASC to CASC. The mode indicated after I, as in I-AUTO, is called pending mode. For example, I-AUTO means the primary controller is presently initialized with AUTO mode pending.

CONFIG
Status variable, defining initialization and/or PV-tracking.

INIT initializes an upstream primary controller's OP if the initializing controller (secondary) is not in CASC mode.
TRACKING forces SPE to assume (track) the value of PVE if the controller is not in AUTO or CASC mode.
I and TR performs both initialization and PV-tracking.

EQUATION
Status variable, defining different selections of inputs for PID control.

Type A calculates PID-control on ERR (ERR = PVE – SPE).
Type B calculates PI-control on ERR (ERR = PVE – SPE) and D-control on PVE only.
Type C calculates I-control on ERR (ERR = PVE – SPE) and PD-control on PVE only.

D.2   Process simulation variables

The parameters used for process simulation are unique to the process simulated. Knowledge of the simulation is not required to complete the exercises in this documentation. It is however recommended to study the *application PCF* file if a deeper understanding is desired.

As most simulations contain noise and disturbance, the following is a description of them, using their generic names.

K-NOISE	Gain of the noise simulation. The values of a random number generator will be superimposed on the simulated process variable. The magnitude of the random numbers is rescaled by K-NOISE.
K-DISTURB	Gain of the automatic random disturbance generator. Internally, within the disturbance generator, a raw disturbance value (RAW-DIST) is calculated by incrementing random numbers (positive and negative). This raw value has to go through a low-pass filter with the time constant TC-DISTURB. The output of the low-pass filter is DISTURB.
TC-DISTURB	This time constant controls the speed (frequency) behavior of DISTURB.
DISTURB	Value of the process disturbance. For manual control of DISTURB as with step functions, set K-DISTURB to 0 and TC-DISTURB to 999.
DIST-HI	Disturbance high limit. DISTURB will not exceed DIST-HI when driven by the automatic random disturbance generator. DIST-HI does not limit manual changes of DISTURB.
DIST-LO	Disturbance low limit. DISTURB will not go below DIST-LO when driven by the automatic random disturbance generator. DIST-LO does not limit manual changes of DISTURB.
SUM-DIST	The sum of disturbance and process simulation.

Appendix E

Configuring a tuning exercise
in a controller

E.1 Exercise using a Honeywell controller – configuring a tuning exercise in a basic controller

Configure the five algorithms shown in Figure E.1, which include control and process simulation. The simulation consists of the tags SIM-FLOW, SIM1, SIM2 and SIM3. SIM-FLOW simulates the fuel flow of the feed heater and SIM1, SIM2 and SIM3 simulate the feed heater's outlet temperature. This simulation is very simplified, but it has stability problems as they would occur in reality. The control is cascade control and consists of the primary controller TC-03 (temperature control) and the secondary controller FC-04 (fuel feed control).

The following data assumes the use of a Honeywell basic controller. It should not be too difficult to implement the same configuration in any other controller.

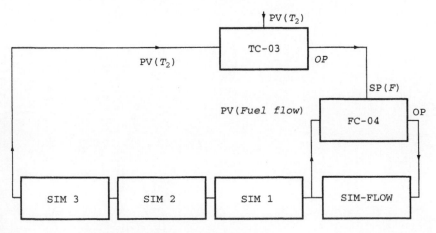

Figure E.1
Block diagram of simplified feed heater simulation and control

	SLOT 3	SLOT 4	SLOT 5	SLOT 6	SLOT 7	SLOT 8
	TC-03	TC-04	SIM-FLOW	SIM1	SIM2	SIM3
CONF	0101	0103	2000	2000	2000	2000
CONF HI	8033	5030	3053	5063	6073	7083
CONF LO	1001	1001	1001	1001	1001	1001
CONF HI LOW	1010	0010	0000	0000	0000	0000
RANGE 0%	0.0	0.0	0.0	0.0	0.0	0.0
RANGE 100%	500.0	100.0	100.0	100.0	100.0	100.0
MODE	MAN	CASC	AUTO	AUTO	AUTO	AUTO
OUT HI%	90.0	100.0	MAX	MAX	MAX	MAX
OUT LO%	10.0	0.0	MIN	MIN	MIN	MIN
PV HI	450.0	90.0	MAX	MAX	MAX	MAX
PV LO	50.0	0.0	MIN	MIN	MIN	MIN
K	?	?	1.0	1.0	1.0	1.0
T1	?	?	0.3	0.3	0.3	2.0
T2	?	?	0.0 ?	0.0 ?	0.0 ?	0.0 ?
TD	0.0 ?	0.0 ?	0.0 ?	0.0 ?	0.0 ?	0.0 ?
0.0 ?	Use the value closest to 0.00 that the controller permits					
MAX	Use the largest value that the controller permits					
MIN	Use the smallest value that the controller permits					
?	You find the best tuning by yourself					

Table E.1
Controller/slot configuration parameters

Appendix F

Installation of simulation software

F.1   Hardware requirements

This document is delivered with a simulation software package. The software is supplied on a 3.5 in. floppy disk.

F.1.1   DOS-based version

This software package requires an IBM compatible computer and an EGA or VGA-graphics card with a minimum of 256 k memory. This software package makes full use of the 256 k graphics memory. Unpredictable results can be expected if the VGA or EGA-graphics card has less memory than required.

This software operates in real time and requires a fast computer system. If the computer is not able to finish all calculations within a given time interval (*scan* time), the message '*slow scan*' will appear on the screen. If the message '*slow scan*' is on the screen, the computer executes in slow motion instead of real time. If the '*slow scan*' message disappears, the computer has regained lost time and resumed real time processing. This software, with its present configuration for training applications (exercises), has been tested on different IBM compatible computers. Most IBM-AT computers operate successfully with 1 s *scan* time (default). (Most IBM-XT compatible computers require a *scan* time of 2 s or longer to execute in real time.) IBM-AT compatible computers with a 16-bit-bus graphics adapter are capable of operating with *scan* times as low as 0.4 s. It is important to note, that even on the slowest computer, all provided exercises run satisfactorily. For example, if a slow computer runs with a *scan* time of 0.5 s, an identical trend display will appear as with fast Pentium machines. The only difference would be the '*slow scan*' message and slow motion execution.

For reasons of time optimization, this software package makes no disk access during execution, except for the save screen function. Therefore, the speed of any disk drive or hard disk is irrelevant. Nevertheless, the software may not run if the disk is write protected. This enables displays to be saved on disk for later retrieval.

The free conventional memory required to run this software package is 450 k. Computers with 640 k of conventional memory may use up to 190 k (640 k = 190 k + 450 k) for DOS and TSRs (terminate and stay resident). There may not be enough memory available if large memory consuming TSR-type programs have been loaded. If you have a problem with inadequate free memory, you will need to remove some TSRs from memory.

F.1.2   Software requirements

This software package runs under MS-DOS version 5.1 or later. The background graphics has the same structure as PCX-files produced by the Microsoft Paintbrush program. A copy of the Paintbrush program is thus required for configuration work.

F.1.3   Installation procedure

Together with the software package comes a text file called README.TXT which explains last minute changes to the installation procedure.

For installation on hard disk or floppy, simply type INSTALL at the DOS-prompt:

A:INSTALL

F.1.4   80 × 87 Auto-detection

The program has in-built auto-detection logic in order to detect the existence of an 80 × 87-co-processor. If an 80 × 87 is available, the program will use it automatically, and if not, the program will emulate the co-processor's functions. Very few IBM compatible PCs return incorrect information, saying that a non-existent 80 × 87 is available, or vice versa. For those situations, where the program cannot detect the presence of a co-processor, the program provides an option for over-riding the auto-detection logic; this option is the 87 environment variable. Refer to your DOS user's manual to find out more about environment variables.

Only if your computer is not capable of returning correct information about the existence of an 80 × 87, set the 87 environment variable at the DOS prompt with the SET command, like this:

C > SET 87 = N

This would advise the program that there is no co-processor.
or like this:

C > SET 87 = Y

This would advise the program that there is a co-processor.

Do not include spaces to either side of the = sign. (If you SET 87 = Y when, in actual fact, there is no 80 × 87 available on the computer system, your computer will lock up.)

F.2   Industrial control in practice – training version 1.1 ICT 32

F.2.1   Introduction

Industrial Control in Practice Training ICT 32 has been developed as an upgrade product to the DOS-based one and is intended only for Windows 95 and Windows NT platforms. The outline below gives you further instructions on how to install the package. Should you require further assistance, you should refer to the accompanying workshop manual *Practical Process Control for Engineers and Technicians* or contact one of our offices listed on this sheet. Alternatively, email us at idc@idc-online.com and put in your subject heading 'ICT-32 software support'. One of our engineers shall endeavor to get back to you as soon as possible.

The program requires a minimum of 16 MB of hard disk space.

CD-ROM Supplied

You should have a CD-ROM clearly labeled:

Industrial Control in Practice

Training Version 1.1 ICT32

From Windows 95 or NT run the program ICT32IDC.EXE

For further assistance:

http://www.idc-online.com

idc@idc-online.com

Check for any damage in shipping such as cracks on the surface. If in doubt contact your nearest IDC office.

F.2.2 Installation instructions

Herewith the instructions for installing the process control software and the Windows link option:

1. Place the disk in your CD drive.
2. From Windows 95 (and NT) RUN the program ICT32IDC.EXE.
3. Enter your own directory name, or use the default directory C:\ICT32IDC.
4. Upon decompression being completed, the program will run, and request:

   (a) Your user name
       Type in *Instrument Data Communications*
   (b) Your user ID Number
       Type over the displayed zero with the number *2146223116*.

5. The program is now installed and ready for use.
6. Create a Windows shortcut, pointing to the directory name you used and the program name ICT32IDC.EXE.
7. Use the standard Windows procedure to generate your own HELP index. (Refer to your Windows applications instructions for guidance on 'FIND Setup Wizard' which relates to this subject.)

F.2.3 Additional software

- The program ICT32I~8.EXE contains 14 'Windows Shortcuts', one for each of the applications contained in the main program, ICT32IDC.EXE.
- These are for use in making up your own 'Windows Folder' for the main program.
- From Windows 95 (and NT) RUN the program ICT32~8.EXE and follow the instructions as they are displayed on your screen.

Note: For the shortcuts to 'work', the main program *must* reside in the default directory named C:\ICT32IDC.

F.2.4 Further notes

The program requires a minimum of 16 MB of hard drive disk space. Upon installation being successfully completed and correctly enabled, the directory must contain a minimum of 37 files (excluding any shortcut files, suffixed with .LNK).

Appendix G

Operation of simulation software

The operation of this training simulation is kept as simple as possible. An operator or student may call up displays, change variables or toggle trend pens. Every other operation totally depends on the configuration of this software package. Therefore, the basic operation is explained first, followed by some operator advice which relates to the training applications delivered.

G.1   Starting the program from DOS

Before you actually start the software, make sure the current directory is the directory where the software has been installed. If this is not the case, use the DOS command CD to change to the directory to location of the software.

C > CD C:\CTL_APPL

Then, type 'MENU', and a menu showing available exercises will appear. Then, use the cursor keys to move the cursor on screen to the appropriate application. Press Enter, and the exercise pointed to, will come up.

G.2   Starting a customized application from the DOS prompt

This is only necessary when starting a custom-made application that is not found in the menu. For each training application there exists a configuration file with the *extension PCF* (process configuration file). The configuration file defines displays, operator access and calculations for control and simulation. The program *CONTROL.EXE* reads the configuration file *filename.PCF* and executes the process as configured.
Start program CONTROL.EXE from the DOS prompt:

DOS prompt > CONTROL filename

An example of starting a customized application is the file LEVEL.PCF:

C > CONTROL LEVEL

After the program name CONTROL, the filename of the process configuration file has to be entered as a command line parameter. This loads and starts the system. The extension PCF is optional.

G.3   Starting the program from MS-Windows

Open the folder with your applications. Select and open the training application you want to run.

G.4   Display call up

The function keys F1 to F8 are designed to call up to eight different displays. The type and the contents of the various displays depend on their associated configuration (PCF) file. There is only one exception – function key F4 is reserved for trend displays only. The contents of the trend display, of course, depend on its configuration.

G.5   How to quit an individual training application

To quit a training application, press simultaneously the keyboard keys *Ctrl* and *Q* or press simultaneously the keys *Ctrl* and *E*. The BREAK-function (Ctrl-C or Ctrl-Break) works, although its use is not recommended.

G.6   Value change of variables

To change the value of a particular variable, type the name of the variable followed by *Enter*. The variable name will appear on the bottom of the screen. Then, type the new value followed by *Enter*. In a few rare cases, the variable name shown on display is too large to be displayed in full. If this is the case, the name may be truncated, but the full variable name is still required to be entered to make changes.

G.7   Command line parameters

Invoke **CONTROL / ?** from the DOS prompt to obtain an explanation of the various command line parameters.

G.8   Status change of status variables

To change the status of a particular status variable (examples: MODE, ALARM, etc.), type the name of the variable followed by *Enter*. The variable name and its current status will appear on the bottom of screen. Then, use the [✱] or [✱] keys to change the status to the higher or lower status value. Otherwise, the same rules apply as described under value change of variables discussed previously.

G.9   Trend operation

The trend display (F4) can have up to eight trend pens, numbered 0–7. The trend pen number assigned to a particular variable depends on its configuration. The actual trending of each pen can be controlled by typing *TREND0*, *TREND1*, *TREND2*, ... , *TREND7*. These commands will terminate the trending of pens which are active and will initiate the trending of pens which are inactive.

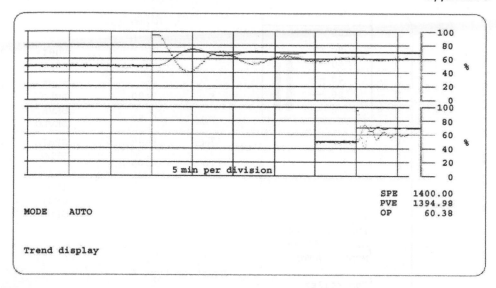

Figure G.1
Trend display

G.10   Display save

Each display can be saved onto disk in the current directory using the commands *SAVE0*, *SAVE1*, *SAVE2*, ... , *SAVE7*. These commands create files named SAVE0.SNP, SAVE1.SNP, SAVE2.SNP, ... , SAVE7.SNP. A saved display can be called upon display F1 only. To call it up, type the number of the display followed by *Enter*.

Note: Graphics capture programs, provided with other software packages, may not capture the graphics actually seen on screen. Most graphics capture programs will capture the first display page (F1) only, although another page is actually on screen. Therefore, it is important to use *SAVE#* to save a screen image and restore it on the first display page (F1), before using a graphics capture program.

G.11   Display assignments as delivered

The training applications (or exercises), supplied with this package, have been configured with the following overall approach:

- The first display page (F1) holds a general cover page, which is the same for all exercises. This cover page is not used for any variables (but could be configured with variables). The display is mainly reserved for displaying previously saved displays. Display F1 also shows the use of the function keys F1 to F12.
- The second display page (F2) shows a block diagram of the particular training application. This display includes the important parameters which the operator will access.
- There are as many detail displays in a training application as there are controllers and major control units. The function keys assigned to each detail display are seen in the first display (F1). Each detail display is designed as an operator interface to control one controller. It displays and permits operations on variables, parameters and limits. The detail displays in the training applications use a similar layout to displays of industrial operator stations.

See Appendix C for further information.

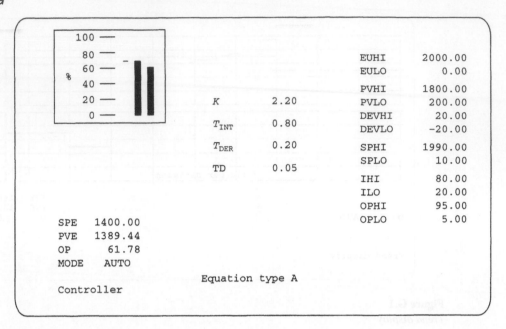

Figure G.2
Detail display

- Display page (F4) always shows a trend display. In addition to the trend's specific commands TREND0, TREND1, etc., normal variable changes can be made on variables displayed. The time scale on the trend display is always in minutes. For every minute, a vertical scaling bar will appear on the fast trend and for every five minutes, a vertical scaling bar will appear on the slow trend, independent of the SCAN time chosen.
- Display page (F8) holds an auxiliary display with all variables necessary for process simulation and control. It also shows those control variables, normally not shown on real industrial control displays, that are helpful in the understanding of process control. This display shows variables such as simulation gain and values, noise gain and values, disturbance gain and values, special controller output and limit calculations, etc. Refer to Appendix D for an explanation of the most frequently used parameters in this display.

Appendix H

Configuration

Configuration is the way of setting up displays, operator access and calculations for control and simulation. The configuration has to be written in a file with the extension PCF (process configuration file). The program *CONTROL.EXE* reads the configuration file *filename.PCF* and executes the process as configured.

Errors within the process configuration file will cause the program CONTROL.EXE to abort and display an error message on screen. Such an error message always contains a file index, pointing to the character within the process configuration file where the error has been detected.

Appendix I

General syntax of configuration commands

Every command within a PCF file starts with '*' and ends with ';'. Everything following a ';' character and before a '*' character is ignored and is therefore useable for comments. Spaces separate different command elements (like parameters). Therefore, no space character is permitted within the elements of a command (except for the *DISPLAY#.DESC: displayname* command's actual descriptor. See *DISPLAY#.DESC: displayname* for a detailed description).

Depending on the type of command, the ':' character is followed by a function or display descriptor (without spaces). Then, parameters follow as explained in the following Chapter 'Configuration commands'. Any parameters in [ ] brackets are optional. The commands have to be written in upper or lower case as shown in the following Chapter 'Configuration commands'.

If a command allows a color to be specified, the following colors or abbreviations (in brackets) can be used: BLACK (KB), BLUE (BL), GREEN (GN), CYAN (C), RED (R), MAGENTA (M), BROWN (BR), WHITE (W), GRAY (GY), HI-BLUE (HBL), HI-GREEN (HGN), HI-CYAN (HC), HI-RED (HR), HI-MAGENTA (HM), HI-YELLOW (HY), HI-WHITE (HW). The color names can be written in upper or lower case.

Appendix J

Configuration commands

J.1 General purpose variable definitions

There are 64 general purpose integer variables, 256 general purpose floating point variables and 64 general purpose status variables. These general purpose variables exist and can be used by algorithms without variable definition. The purpose of the variable definition process is to attach variable names and initialization values to general purpose variables.

Names have to be attached to variables, if they have to be displayed and if operator access is required. Initialization values will be used at start-up time only to set variables to their initial values.

- *INT#:name [v = nnnn];

  This defines integer variables. The use is totally open to the individual application.

#	Number of the integer to be defined (0–63). Undefined integers may be used in algorithms, but will not be initialized and have no name attached.
name	Text string to be used as integer descriptor (maximum of 10 characters).
v	Initialization value set to nnnn (0 to ±9999).

Example:

```
*INT23:BUFFIDX          v = 132;deadtime buffer: float132 to 153 minutes
*FLOAT34:DEADTIME       v = 0.3;
*FLOAT35:INPUT          v = 50;
*FLOAT83:OUTPUT         v = 50;
*ALGO5:DEADTIME         P1 = 83   P2 = 35   P3 = 34   P4 = 23   P5 = 24;
```

The parameters P4 and P5 of the *deadtime* algorithm have to be of integer type. Their purpose is to provide pointers for calculating history data. Parameter P4 contains the index number of the first of 22 sequential floating-point variables, to be used as deadtime history buffer. Parameter P5 contains the index number of the floating-point variable, containing the oldest history value. No parameter definition for integer 24 is required, since integer 24 is used as an internal parameter of *deadtime* and neither initialization nor display is necessary.

- *STATUS#:name [v = nnnn] [s0 = string] [s2 = string] ... [s7 = string];
  This defines status variables. The use is totally open to the individual application.

#	Number of the status variable to be defined (0–63). Undefined status variables may be used in algorithms, but will not be initialized and have no name attached.
name	Text string to be used as variable descriptor (maximum of 10 characters).
v	Initialization value set to nnnn (0–7).
s0–s7	Status descriptors to be used for display of up to eight status strings depending on the value of the status variable (s0 for value of 0, s1 for value of 1 up to s6 for value of 6. s7 is the default for all other values). No regular space character is permitted within the descriptor and all status descriptors have to be of equal length. The character 'Ç' has to be used, wherever a *space* character is needed. 'Ç' will be used like a *space* character. The character 'Ç' is obtained by holding down the Alt key, while sequentially pressing the keys 1 2 8 on the numeric keypad.

Example:

*STATUS6:PVALARM	v = 0	s0 = ÇÇÇÇÇÇÇÇ	s1 = INSTÇHIÇ	s2 = INSTÇLOÇ
		s3 = ÇROCÇHIÇ	s4 = ÇROCÇLOÇ	
		s5 = ÇPVÇHIÇÇ		
		s6 = ÇPVÇLOÇÇ	s7 = ÇDEVÇHIÇ	
		s8 = ÇDEVÇLOÇ;		
*STATUS7:OPALARM1	v = 0	s0 = ÇÇÇÇÇÇÇÇ	s1 = ÇOPÇHIÇÇ	
		s2 = ÇOPÇLOÇÇ	s3 = ÇINTÇHIÇ	
		s4 = ÇINTÇLOÇ;		
*INT1:STATUS1	v = 0;			
*ALGO5:ALARM	P1 = 6	P2 = 7 P3=1;		
*DISPLAY2.UPDATE:stv6	col = 6		row = 17	fg = HI-RED
	bg = BLACK;			
*DISPLAY2.UPDATE:stv7	col = 6		row = 22	fg = HI-RED
	bg = BLACK;			

An integer variable is used as a general status variable in PID controllers as described in the Chapter 'Algorithms'. In our example, integer 1, which has been given the name STATUS1, stores such controller status. Among other status information, integer 1 stores the total alarm status of the whole PID controller. To make status variables available for display configuration, the algorithm ALARM drives two status variables, status 6 (PVALARM1) and status 7 (OPALARM1).

- *FLOAT#:name [v = nnnn];
  This defines float variables. The use is totally open to the individual application.

#	Number of the float variable to be defined (0–255). Undefined float variables may be used, but will not be initialized and have neither name nor extended definitions attached.
name	Text string to be used as a float variable descriptor (maximum of 10 characters).
v	Initialization value set to nnnn (nnnn from 0.00 to ±9999.90).

Example:

```
*FLOAT32:LAG-INPUT      v = 20;
*FLOAT11:LAG-OUTPUT     v = 20;
*FLOAT97:LAG-TC         v = 0.5;
*ALGO20:LAG             P1 = 11   P2 = 32   P3 = 97;
```

All three floating point variables used by the LAG algorithm have been given names and start-up values.

J.2   Extended variable definitions

Extended definitions for trend pens (trend display) can be made for, up to eight floating point variables. Similarly, extended definitions for general bar graphs can be made for, up to eight floating point variables. The floating point variables extended for trend pens can be different from those extended for general bar graphs. However, there is no requirement that they have to be different.

- *HISTORY#:fv## [tmin = nn][tmax = mm][c = Color];
  This command assigns a float variable for history collection and activates history collection. *Display 4 is fixed for trend.* Trend occupies text rows 1–14. Rows 0 and 15–24 are available for normal update.

#	Trend number (from 0 to 7) for which the command is valid.
fv##	Variable (float) to be assigned to history collection for trend #.
Tmin = nn	Defines the bottom value (0% value) for trend and bar displays.
Tmax = mm	Defines the top value (100% value) for trend and bar displays.
C = Color	Defines the color for trend displays of the defined variable.

Example:

```
*FLOAT32:LAG–INPUT     v = 20;
*FLOAT11:LAG–OUTPUT    v = 20;
*FLOAT97:LAG–TC        v = 0.5;
*ALGO20:LAG            P1 = 11      P2 = 32      P3 = 97;
*HISTORY3:fv32         tmin = − 20  tmax = 260   c = CYAN
*HISTORY7:fv11         tmin = − 20  tmax = 260   c = CYAN
```

All three floating point variables used by the LAG algorithm have been defined first. Only input and output variables of the LAG algorithm have extended definitions for trending purposes. They have been assigned to trend pens 3 and 7. The trends will be displayed on display 3 (trend display) without further display configuration.

- *LEVEL#:name [v = nnnn] [c1 = cc] [r1 = rr] [c2 = cc] [r2 = rr];
  This defines level variables of float type. The use is totally open to the individual application. The display appears as a bar graph with the two diagonal corners c1 – r1 and c2 – r2.

#	Number of level variable to be defined (from 0 to 7).
name	Text string to be used as variable descriptor (maximum of 10 char).
v	Initialization value set to nnnn (nnnn from 0.00 to +100.00). The range is fixed to 0.00 = 0% bar graph and 100.00 = 100% bar graph.
c1 r1 c2 r2	Position parameters, representing the two diagonal corners of the level update area, within which a vertical bar graph update takes place.

Example:

*FLOAT9:TANKLEVEL	v = 50;	
*LEVEL3:TANKLEVEL	v = 50	c1 = 6 r1 = 5 c2 = 30 r2 = 15;
*DISPLAY1.UPDATE:lv3	fg = HI-BLUE bg = BLACK;	
*DISPLAY5.UPDATE:lv3	fg = HI-BLUE bg = BLACK;	

The floating point variable 9 has been named TANKLEVEL and has been given extended definition as variable number 3 of type level with a defined area on screen. The display update commands DISPLAY1 and DISPLAY5 define the variable level 3 to be displayed in both, display 1 and 5.

J.3   Display commands

All display commands, with the exception of command '*DISPLAY#.BACKGRND: filename;', provide position and color definitions. The existence of these options is shown in the command descriptions as follows:

- ..... [• • • •]
  Format of these definitions:
  ..... [col = cc][row = rr][fg = colorname][bg = colorname]

col = cc	Char column on screen (cc from 0 to 39), where descriptor or update starts (left).
row = rr	Char row on screen (rr from 0 to 23) of descriptor or update.
fg = colorname	Foreground color definition.
bg = colorname	Background or character box color definition.
colorname	The following names are available as color definitions: BLACK (BK), BLUE (BL), GREEN (GN), CYAN (C), RED (R), MAGENTA (M), BROWN (BR), WHITE (W), GRAY (GY), HI-BLUE (HBL), HI-GREEN (HGN), HI-CYAN (HC), HI-RED (HR), HI-MAGENTA (HM), HI- YELLOW (HY), HI-WHITE (HW)

Example:

```
*FLOAT23:PRESSURE          v = 45;
*DISPLAY1.UPDATE:fdesc23   col = 12     row = 5  fg = HI-WHITE
                           bg = BLUE;
*DISPLAY1.UPDATE:fv23      col = 20     row = 5  fg = HI-WHITE
                           bg = BLUE;
```

PRESSURE is the name of floating point variable number 23 and will be displayed in display 1, text row 5, starting at column 12. The actual value of floating point variable 23 will be displayed and updated continuously in display 1, text row 5, starting at text column 20.

- *DISPLAY#.BACKGRND:filename;
This command defines the file containing the background graphics data to be displayed.

  #   Display number (0–7) for which the command is valid. If no DISPLAY#.BACKGRND command exists, the display is considered as non-existent by the system.
  filename filename.EGA contains the background graphics data of display #. If filename.EGA does not exist, a black background will be used. See the Chapter 'Background design' as well.

Example:

*DISPLAY6.BACKGRND:project;

If a graphics data file of the name PROJECT.EGA exists, the file contents will be used as background graphics for display 6. If no graphics data file of the name PROJECT.EGA exists, the background of display 6 will be black. In both cases, display 6 is configured and useable (active).

- *DISPLAY#.DESC:displayname, [• • • •];

  #   Display number (0–7) for which the command is valid.
  displayname Description of display #. The string must not have more than 30 characters and must end with a ',' as a parameter separator. The ',' is not part of the descriptor any more. This way, *space* characters are permitted within the descriptor itself and are not considered as end of descriptor.

Example:

```
*DISPLAY3.DESC:Tuning Trend,   col = 3       row = 24
                               fg = HI-GREEN bg = BLACK;
```

Display 3 will display the title 'Tuning Trend' in green on black background, in text row 24, starting at text column 3.

- *DISPLAY#.UPDATE:fv## [• • • •];
This defines a float variable to be updated at a position of column and row.

  #   Display number (0–7) for which the command is valid.
  fv## Float variable to be displayed and continuously updated.

- *DISPLAY#.UPDATE:fdesc## [• • • •];

This defines a float variable descriptor to be displayed at a position of column and row.

| # | Display number (0–7) for which the command is valid. |
| Fdesc## | Float variable descriptor to be displayed. Update of descriptor takes place at display call-up only. |

Example:

```
*FLOAT23:PRESSURE              v = 45;
*DISPLAY1.UPDATE:fdesc23    col = 12    row = 5    fg = HI-WHITE    bg = BLUE;
*DISPLAY1.UPDATE:fv23       col = 20    row = 5    fg = HI-WHITE    bg = BLUE;
```

PRESSURE is the name of floating point variable 23 and will be displayed in display 1, text row 5, starting at column 12. The actual value of floating point variable 23 will be displayed and updated continuously in display 1, text row 5, starting at text column 20.

- *DISPLAY#.UPDATE:iv## [• • • •];

This defines an integer variable to be updated at a position of column and row.

| # | Display number (0–7) for which the command is valid. |
| iv## | Integer variable to be displayed and continuously updated. |

- *DISPLAY#.UPDATE:idesc## [• • • •];

This defines an integer variable descriptor to be displayed at a position of column and row.

| # | Display number (0–7) for which the command is valid. |
| idesc## | Integer variable descriptor to be displayed. Update of descriptor takes place at display call-up only. |

Example:

```
*INT2:COUNTER                 v = 45;
*DISPLAY7.UPDATE:idesc2    col = 12   row = 5   fg = HI-WHITE    bg = BLUE;
*DISPLAY7.UPDATE:iv2       col = 20   row = 5   fg = HI-WHITE    bg = BLUE;
```

COUNTER is the name of integer variable 2 and will be displayed in display 7, text row 5, starting at column 12. The actual value of integer variable 2 will be displayed and updated continuously in display 7, text row 5, starting at text column 20.

- *DISPLAY#.UPDATE:stv## [• • • •];

This defines a status variable to be displayed and continuously updated at a position of column and row.

| # | Display number (0–7) for which the command is valid. |
| stv## | Status variable to be displayed and continuously updated. |

- *DISPLAY#.UPDATE:stdesc## [• • • •];

This defines a status variable descriptor to be displayed at a position of column and row.

#	Display number (0–7) for which the command is valid.
Stdesc##	Status variable descriptor to be displayed. Update of descriptor takes place at display call-up only.

Example:

*STATUS5:ALARM	v = 1;			
*DISPLAY6.UPDATE:stdesc5	col = 12	row = 5	fg = HI-WHITE	bg = BLUE;
*DISPLAY6.UPDATE:stv5	col = 20	row = 5	fg = HI-WHITE	bg = BLUE;

ALARM is the name of status variable 5 and will be displayed in display 6, text row 5, starting at column 12. The actual status of status variable 5 will be displayed and updated continuously in display 6, text row 5, starting at text column 20.

- *DISPLAY#.UPDATE:lv## [• • • •];
  This defines a level variable to be displayed and continuously updated. The position of column and row in this command will be ignored, since the definition of the '*LEVEL#' command already defines the position. The position of a level variable is the same for each display it is configured.

#	Display number (0–7) for which the command is valid.
lv##	Level variable to be displayed and updated.

- *DISPLAY#.UPDATE:ldesc## [• • • •];
  This defines a level variable descriptor to be displayed at a position of column and row.

#	Display number (0 to 7) for which the command is valid.
idesc##	Integer variable descriptor to be displayed. Update of descriptor takes place at display call-up only.

Example:

*FLOAT9:TANKLEVEL	v = 50;				
*LEVEL3:TANKLEVEL	v = 50	c1 = 6	r1 = 5	c2 = 30	r2 = 15;
*DISPLAY1.UPDATE:lv3	fg = HI-BLUE		bg = BLACK;		
*DISPLAY5.UPDATE:lv3	fg = HI-BLUE		bg = BLACK;		
*DISPLAY5.UPDATE:ldesc3	col = 6	row = 17	fg = HI-GREEN	bg = BLUE	

The floating point variable 9 has been named TANKLEVEL and has been given extended definition as variable number 3 of type level with a defined area on screen. The display update commands DISPLAY1 and DISPLAY5 define the variable level 3 to be displayed in both, display 1 and 5. The descriptor of variable level 3 is displayed in display 5 only.

- *DISPLAY#.BAR:## [• • • •];
  Only those parameters which were assigned for history and trend can be displayed as bar graphs. Bar graphs have a height of 100 pixels and a width of 3 pixels. The update overwrites text, overlapping the bar update area. The bar update area is one character wide and 8 characters high (rows 1–8).

#	Display number (0–7) for which the command is valid.
##	History (Trend) number (0–7) for which the command is valid.

Note: Row definition is fixed. Any row definition will be ignored.

Example:

```
*FLOAT11:PV        v = 250;
*HISTORY7:fv11     tmin = 0     tmax = 500     c = CYAN
*DISPLAY2.BAR:7    col = 6      fg = HI-CYAN   bg = BLACK;
```

Floating point variable 11 has been designed as trend pen 7. The initialization value of 250 represents 50% of range for trend and bar graph as well. Bar graph update takes place in display 2 column 6.

- *DISPLAY#.MARKER:## [• • • •];
  Only those parameters which were assigned for history and trend can be displayed as markers. Markers occupy a height of 100 pixels and a width of 3 pixels. The update overwrites text, overlapping the marker update area. The marker update area is one character wide and 8 characters high (rows 1–8).

#	Display number (0–7) for which the command is valid.
##	History (trend) number (0–7) for which the command is valid.

Note: Row definition is fixed. Any other row definition will be ignored.

Example:

```
*FLOAT10:SP          v = 250;
*HISTORY7:fv11       tmin = 0    tmax = 500     c = CYAN
*DISPLAY2.MARKER:7   col = 4     fg = HI-CYAN   bg = BLACK;
```

Floating point variable 11 has been designed as trend pen 7. The initialization value of 250 represents 50% of range for trend and marker as well. Marker update takes place in display 2 column 4.

J.4   Algorithm definitions

Algorithm definitions are used to define the algorithms their sequence of execution. The algorithms make use of general purpose variables as explained in the Chapter 'Algorithms'.

- *ALGO#:algoname P1 = vnbr P2 = vnbr P3 = vnbr ... Pn = vnbr;
  The ALGO command defines the algorithm to be calculated. There is no direct association between any algorithm and tags or displays. It is the task of the person configuring the system to keep the algorithms organized properly. The sequence of calculations within each scan is determined by the digits which make up the last characters of the ALGO# command. (Example: ALGO5 is the algorithm calculated fifth.)

#	Sequence number defining when the algorithm is calculated in sequence. The maximum number of algorithms for the whole system is 128. Therefore '#' can be from 0 to 127.

algoname   Has to be a legal and proper name of an algorithm supported. A list of supported algorithms is in a separate chapter 'Algorithms'.

P1 to Pn   The number of parameters is dependent on the type of algorithm chosen. vnbr defines which variable has to be passed on to the algorithm. There is no need to specify whether it is an integer or float variable, the algorithm picks the right one.

Note: Be careful and know, whether the algorithm is using an integer, an integer with status strings or a float variable.

Appendix K

Algorithms

The following is a list of Algorithms with the ability to interact automatically with each other to provide a complete controller. Depending on the combination of these algorithms and their specific use, controllers different in function and complexity can be configured.

PIDN	Incremental real PID
PIDX	Incremental ideal PID
PV	PV and SP limits, mode, initialization, tracking, etc.
OP	PID-OP value, limits, mode, initialization, etc.
MODE	MODE handling between PID?, PV, OP, ALARM, etc.
STATUS	Controller status
ALARM	Controller alarm

The following is a list of Stand Alone Algorithms which can be used in any combination:

LEADLAG	First order LEAD and LAG
SUM	Summer
MUL	Multiplier
RATIO	Ratio and bias
PROP	Incremental proportional with gain
INT	Integral
DERN	Real derivative
DERX	Ideal derivative
LAG	First order LAG
NOISE	Random noise with gain
DISTURB	General disturbance generator
SINE	Sine wave generator
SNDORDER	2nd order system with variable damping
HILIM	High limit
LOLIM	Low limit
ROCHILIM	Rate of change limit high
ROCLOLIM	Rate of change limit low
ILIM	Integral limits
HIAL	High alarm

LOAL	Low alarm
ROCHIAL	Rate of change alarm high
ROCLOAL	Rate of change alarm low
FREQTOTC	Frequency to time constant conversion
EUTOPCT	Engineering units to % conversion
PCTTOEU	% to engineering units conversion
LINK	Link if a status variable is true
MATH	General mathematical
DIV	Division
HEATCOMP	Heat balance calc for feedforward control
HEATSIM	Heat as function of fuel consumption for simulation
MASSFLOW	Mass-flow compensation
DEADTIME	Deadtime for simulation and/or control

K.1 Interacting PID-algorithm-blocks to build PID-controllers

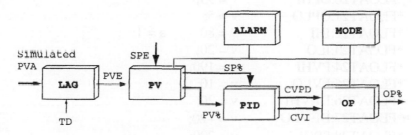

Figure K.1
Principle interaction between algorithms of a PID controller

K.1.1 Configuration example

Level controller algorithms and variables

*ALGO1:LAG	P1 = 14	P2 = 46	P3 = 10;	
*ALGO2:PV	P1 = 14	P2 = 15	P3 = 4	P4 = 2
	P5 = 3	P6 = 1	P7 = 8	P8 = 1
	P9 = 5	P10 = 6	P11 = 24	P12 = 25
	P13 = 26	P14 = 27	P15 = 28	P16 = 29;
*ALGO3:PIDN	P1 = 11	P2 = 12	P3 = 14	P4 = 15
	P5 = 7	P6 = 8	P7 = 9	P8 = 2
	P9 = 16	P10 = 17	P11 = 18;	
*ALGO4:OP	P1 = 1	P2 = 11	P3 = 12	P4 = 1
	P5 = 3	P6 = 4	P7 = 1	P8 = 2
	P9 = 22	P10 = 23	P11 = 20	P12 = 21
	P13 = 13;			
*ALGO5:ALARM	P1 = 6	P2 = 7	P3 = 1;	
*ALGO6:MODE	P1 = 5	P2 = 1	P3 = 1	P4 = 2
	P5 = 8;			

```
*FLOAT1:OP1           v = 50;
*FLOAT2:PVE1          v = 100;
*FLOAT3:SPE1          v = 100;
*FLOAT4:CSP1          v = 100;
*FLOAT5:EUHI          v = 200;
*FLOAT6:EULO          v = 0;
*FLOAT7:K             v = 3;
*FLOAT8:T_INT         v = 1;
*FLOAT9:T_DER         v = 0;
*FLOAT10:TD           v = 0;
*FLOAT11:CVPD         v = 0;
*FLOAT12:CVI          v = 0;
*FLOAT13:OPVIRT1      v = 50;
*FLOAT14:PV           v = 50;      %
*FLOAT15:SP           v = 50;      %
*FLOAT16:LASTD        v = 0;
*FLOAT17:LASTP        v = 0;
*FLOAT18:LRATE        v = 0;       Used in PIDX algorithm only!
*FLOAT20:OPHI         v = 95;
*FLOAT21:OPLO         v = 5;
*FLOAT22:IHI          v = 80       a = 1;
*FLOAT23:ILO          v = 20;
*FLOAT24:PVHI         v = 190;
*FLOAT25:PVLO         v = 10;
*FLOAT26:DEVHI        v = 20;
*FLOAT27:DEVLO        v = −20;
*FLOAT28:SPHI         v = 200;
*FLOAT29:SPLO         V = 0;
```

```
*STATUS1:MODE1        v = 1        s0 = ÇMANUALÇ     s1 = ÇÇAUTOÇÇ
s2 = ÇÇCASCÇÇ                      s3 = ÇI-MANÇÇ     s4 = ÇI-AUTOÇ
                                   s5 = ÇI-CASCÇ;
*STATUS2:EQUATION     v = 0        s0 = ÇTYPEÇAÇ     s1 = ÇTYPEÇBÇ
                                   s2 = ÇTYPEÇCÇ;
*STATUS3:ACTION1      v = 1        s0 = ÇDIRECTÇ     s1 = ÇREVERSE;
*STATUS4:OPCALC1      v = 1        s0 = ÇÇREALÇÇ     s1 = ÇÇVIRTÇÇ;
*STATUS5:INIT1        v = 0        s0 = ÇÇÇÇÇÇÇ      s1 = ÇÇINITÇÇ;
*STATUS6:PVALARM1     v = 0        s0 = ÇÇÇÇÇÇÇ      s1 = INSTÇHIÇ
                                   s2 = INSTÇLOÇ
                                   s3 = ÇROCÇHIÇ     s4 = ÇROCÇLOÇ
                                   s5 = ÇPVÇHIÇÇ     s6 = ÇPVÇLOÇÇ
                                   s7 = ÇDEVÇHIÇ     s8 = ÇDEVÇLOÇ;
*STATUS7:OPALARM1     v = 0        s0 = ÇÇÇÇÇÇÇ      s1 = ÇOPÇHIÇÇ
                                   s2 = ÇOPÇLOÇÇ
                                   s3 = ÇINTÇHIÇ     s4 = ÇINTÇLOÇ;
*STATUS8:CONFIG1      v = 0        s0 = ÇÇÇÇÇÇÇ      s1 = ÇÇINITÇÇ
                                   s2 = TRACKING     s3 = ÇIÇ&ÇTRÇ;
*INT1:STATUS1         v = 0;
*FLOAT46:LVL-SIM      v = 50;
```

K.1.2 Status word for PID-block interaction and general alarms

External status comes from downstream point and has the same structure as the internal status.

Set by and used	Used	Used	Parameter name
MODE	MODE-PRIM		INIT
PV			INSTHI
PV			INSTLO
OP	OP-PRIM	ALARM	IHI
OP	OP-PRIM	ALARM	ILO
OP		ALARM	OPHI
OP			OPLO
PV	OP-PRIM		SPHI
PV	OP-PRIM		SPLO
PV		ALARM	DEVHI
PV		ALARM	DEVLO
		ALARM	ROCHI
		ALARM	ROCLO
PV		ALARM	PVHI
PV		ALARM	PVLO

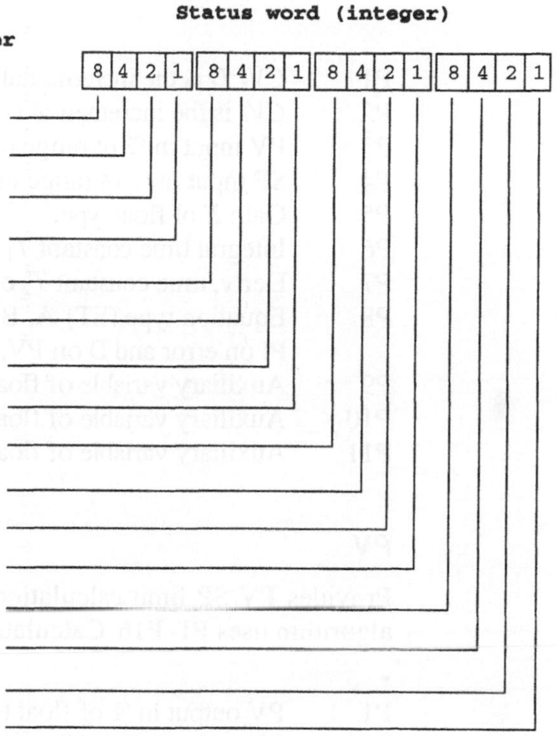

Figure K.2

Status word for PID-block interaction and general alarms

PIDN

PID-NORMAL (real) Algorithm. This algorithm is best suited as a field controller (secondary with field I/O, etc.).

$$K \times (1+sT_2)/(1+saT_2) \times (1+1/sT_1) a = 1/10$$

- P1 *CVPD* is the incremental *c*ontrol *v*alue for *PD* control of float type.
- P2 *CVI* is the incremental *c*ontrol *v*alue for *I* control of float type.
- P3 PV input in % of range of float type.
- P4 SP input in % of range of float type.
- P5 Gain *K* of float type.
- P6 Integral time constant T_1 of float type.
- P7 Deriv. time constant T_2 of float type.
- P8 Equation type (ET) A, B or C of status type. ET = A for PID on error. ET = B for PI on error and D on PV. ET = C for I on error and PD on PV.
- P9 Auxiliary variable of float type to hold previous scan's derivative calculation.
- P10 Auxiliary variable of float type to hold previous scan's proportional calculation.

PIDX

PID-NORMAL (ideal) algorithm. This algorithm is best suited as a high level controller (like computer-resident primary controllers), without direct field I/O, using pre-processed data only.

$$K \times (1 + 1/sT_1 + sT_2)$$

P1	*CVPD* is the incremental *control value* for *PD* control of float type.
P2	*CVI* is the incremental *control value* for *I* control of float type.
P3	PV input in % of range of float type.
P4	SP input in % of range of float type.
P5	Gain *K* of float type.
P6	Integral time constant T_1 of float type.
P7	Deriv. time constant T_2 of float type.
P8	Equation type (ET) A, B or C of status type. ET = A for PID on error. ET = B for PI on error and D on PV. ET = C for I on error and PD on PV.
P9	Auxiliary variable of float type to hold previous scan's derivative calculation.
P10	Auxiliary variable of float type to hold previous scan's proportional calculation.
P11	Auxiliary variable of float type to hold previous scan's rate of change calculation.

PV

Provides PV/SP limit calculation, alarm status setting and conversion to % values. This algorithm uses P1–P16. Calculation should take place before PID.

P1	PV output in % of float type.
P2	SP output in % of float type.
P3	Local SP (LSP) in EU of float type.
P4	PV input in EU of float type.
P5	SP input in EU of float type.
P6	MODE of control (0 = MAN, 1 = AUTO, 2 = CASC, 3 = I-MAN, 4 = I-AUTO, 5 = I-CASC) of status type.
P7	PV-TRACKING (0 = OFF, 1 = ON) of status type.
P8	STATUS-WORD of status type (see explanation before).
P9	EUHI (upper range limit) of float type.
P10	EULO (lower range limit) of float type.
P11	PVHI (PV alarm limit) of float type.
P12	PVLO (PV alarm limit) of float type.
P13	DEVHI (PV-SP alarm limit) of float type.
P14	DEVLO (PV-SP alarm limit) of float type.
P15	SPHI (SP limit) of float type.
P16	SPLO (SP limit) of float type.

OP

Calculates the absolute value based on the increments available from PIDN or PIDX algorithm. Multiple OP-algorithm blocks may be added to the same PID-block if multiple OP destinations are required. Initialization of the OP value takes place automatically if used in cascade mode. Uses P1–P13.

P1   OP (control destination).

P2   *CVPD* is the incremental *control value* for *PD* control of float type.

P3   *CVI* is the incremental *control value* for *I* control of float type.

P4   MODE of control (0 = MAN, 1 = AUTO, 2 = CASC, 3 = I-MAN, 4 = I-AUTO, 5 = I-CASC) of status type.

P5   REV control action if 1 (0 = direct) of status type.

P6   LT is the limit calculation type (0 = real/unsaturated OP-limits, 1 = virtual/saturated OP-limits) of status type.

P7   STATUS-WORD of this controller of status type (see explanation before).

P8   External STATUS-WORD of downstream (secondary) controller of status type (see explanation before).

P9   IHL (integral limit) of float type.

P10   ILL (integral limit) of float type.

P11   OPHI (OP limit) of float type.

P12   OPLO (OP limit) of float type.

P13   OPVIRT is the virtual, internal and unlimited OP value for saturated OP limit calculation of float type. This parameter has to be provided, independently of the value of parameter LT.

MODE

Monitors and calculates Initialized and non-initialized mode values. It serves for the propagation of INIT between primary and secondary controllers.

P1   INIT for display purpose only of status type.

P2   MODE of control (0 = MAN, 1 = AUTO, 2 = CASC, 3 = I-MAN, 4 = I-AUTO, 5 = I-CASC) of status type.

P3   STATUS-WORD of this controller of status type (see previous explanation).

P4   External STATUS-WORD of downstream (secondary) controller of status type (see previous explanation).

P5   INITIALISATION and PV-TRACKING variable of status type.

ALARM

Provides two alarm status Var (PVALARM, OPALARM) for display and general purpose. The two variables are set in accordance with the STATUS-WORD (see explanation before) or any ALARM-STATUS-WORD provided by single alarm algorithms.

P1   PVALARM output (PV alarm status) of status type.

P2   OPALARM output (OP alarm status) of status type.

P3   STATUS-WORD of status type (see explanation before).

K.2   General algorithms

LEAD–LAG

First order lead–lag calculation

$$\frac{(1 + sT_1)}{(1 + sT_2)}$$

P1	Output of float type.
P2	Input of float type.
P3	Lead time constant T_1 of float type.
P4	Lag time constant T_2 of float type.
P5	Auxiliary variable of float type to hold previous scan's lag calculation.

SUM

'Summer' to add two inputs: Output = InputA + InputB

P1	Output of float type.
P2	InputA of float type.
P3	InputB of float type.

MUL

'Multiplier' to multiply two inputs: Output = InputA × InputB

P1	Output of float type.
P2	InputA of float type.
P3	InputB of float type.

RATIO

Ratio calc.: Output = Input × Ratio + Bias

P1	Output of float type.
P2	Input of float type.
P3	Ratio of float type.
P4	Bias of float type.

PROP

Incremental proportional calculation: ¡Output = K × ¡Input

P1	Output of float type.
P2	Input of float type.
P3	K (gain) of float type.
P4	Auxiliary variable of float type to hold previous scan's output calculation.

INT

Incremental integral calculation: $1/sT$

P1	Output of float type.
P2	Input of float type.
P3	T (Integral time constant) of float type.

DERN

Incremental derivative calculation (real form): $(1 + sT)/(1 + saT)a = 1/10$

P1	Output of float type.
P2	Input of float type.
P3	T (Deriv. time constant) of float type.
P4	Auxiliary variable of float type to hold previous scan's derivative calculation.

DERX

Incremental derivative calculation (real form): sT

P1   Output of float type.
P2   Input of float type.
P3 T (Deriv. time constant) of float type.
P4   Auxiliary variable of float type to hold previous scan's derivative calculation.

LAG

First order lag calculation $1/(1 + sT)$

P1   Output of float type.
P2   Input of float type.
P3   Time constant in minutes of float type.

NOISE

Superimposed noise calculation:

$$Output = Input + K \times RANDOM$$

where by RANDOM is noise between –0.5 and +0.5.

P1   Output of float type.
P2   Input of float type.
P3 K (Noise-Gain) of float type.

DISTURB

General disturbance generator

$$RawDist = RawDist + K \times RANDOM$$
$$Output = (LAG \text{ with } TC) \times RawDisturb$$

where LAG serves as low-path filter and as return time constant for manual changes of Output.

P1   Output of disturbance of float type.
P2 K (disturbance-gain) of float type.
P3   TC in minutes and of float type.
P4   Raw disturbance value of float type.
P5   Disturbance HI-limit of float type.
P6   Disturbance LO-limit of float type.
P7   Disturbance BIAS of float type.

SINE

Sine wave generator $w/(s^2 + w^2)$

P1   Output of sine wave of float type (initialize to value of base line).
P2   Baseline of float type.
P3   Time constant in minutes of float type.
P4   Auxiliary variable of float type (initialize to first maximum or first minimum)

SNDORDER

Second order system with variable damping

$$\frac{1}{(s^2 + 2\underline{\Omega}_n s_n{}^2)}$$

$$W_d = W_n (1 - \underline{\Omega} \, d^2)^{1/2}$$

$$Tn = \frac{1}{W_n}$$

P1   Output of float type.
P2   Input of float type.
P3   Damping factor $\underline{\Omega}$ of float type.
P4   Natural time constant Tn in minutes of float type.
P5   Auxiliary variable of float type (initialization value of P2 and P5 should be equal).

HILIM

The value of P1 will not exceed the value of P2.

P1   Variable of float type, to be monitored and clamped at limit.
P2   Limit of P1 of float type.

LOLIM

The value of P1 will not go below the value of P2.

P1   Variable of float type to be monitored and clamped at limit.
P2   Limit of P1 of float type.

ROCHILIM

The value of P1 will not increment by more than the value of P2.

P1   Variable of float type, to be monitored and manipulated by the limit when necessary.
P2   Increment limit of P1 of float type (rate of change per minute).
P3   Auxiliary variable of float type.

ROCLOLIM

The value of P1 will not decrement by more than the value of P2.

P1   Variable of float type, to be monitored and manipulated by the limit when necessary.
P2   Decrement limit of P1 of float type (rate of change per minute).
P3   Auxiliary variable of float type.

ILIM

Incremental integrator with integral limits and integral wind-up handling: $1/sT$

P1   Output of float type.
P2   Input of float type.
P3 T (integral time constant) of float type.

P4 Alarm of status type:
    0 = No integral alarm
    1 = Integral high, increment suspended
    2 = Integral low, decrement suspended.
P5 Integral high limit of float type.
P6 Integral low limit of float type.

HIAL

An alarm will be raised when the value of P2 exceeds the value of P3. P2 is not clamped and can have any value.

P1 Alarm of status type:
    0 = No alarm
    1 = High alarm.
P2 Variable of float type, to be monitored.
P3 Alarm limit of P1 of float type.

LOAL

An alarm will be raised when the value of P2 is below the value of P3. P2 is not clamped and can have any value.

P1 Alarm of status type:
    0 = No alarm
    1 = Low alarm.
P2 Variable of float type, to be monitored.
P3 Alarm limit of P1 of float type.

ROCHIAL

An alarm will be raised when the increment of P2 is above the value of P3. The P2 increment is not limited and can have any value.

P1 Alarm of status type:
    0 = No alarm
    1 = ROC high alarm.
P2 Variable of float type, to be monitored.
P3 Increment alarm limit of P2 of float type (rate of change per minute).
P4 Auxiliary variable of float type.

ROCLOAL

An alarm will be raised when the decrement of P2 is below the value of P3. The P2 decrement is not limited and can have any value.

P1 Alarm of status type:
    0 = No alarm
    1 = ROC low alarm.
P2 Variable of float type, to be monitored.
P3 Decrement alarm limit of P2 of float type (rate of change per minute).
P4 Auxiliary variable of float type.

STATUS

A general status word will be updated, based on alarm status variables, set by algorithms as shown in table below. The format of this general status word is consistent with the PID controller status word.

P1   General status word of integer type.
P2 –P10   Individual variables of status type as shown in table below.

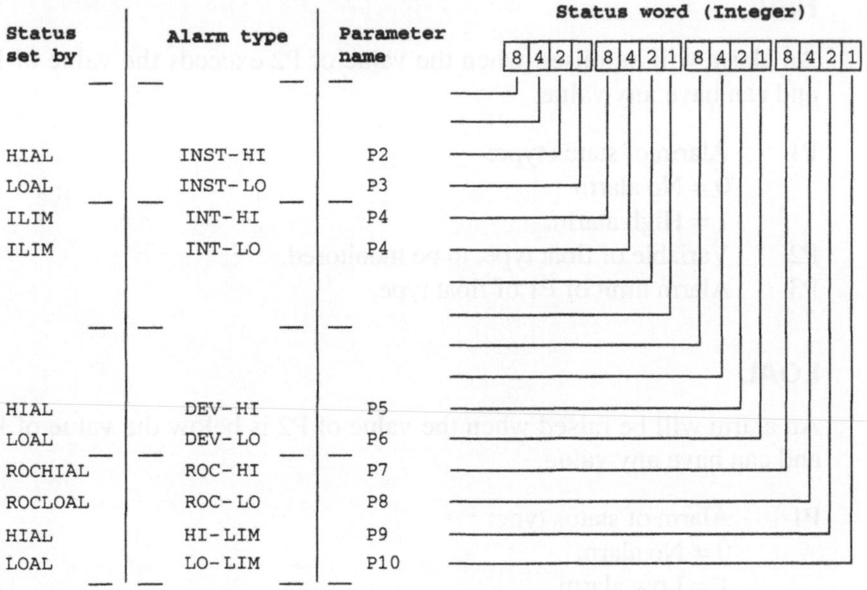

Status set by	Alarm type	Parameter name	Status word (Integer)
HIAL	INST-HI	P2	
LOAL	INST-LO	P3	
ILIM	INT-HI	P4	
ILIM	INT-LO	P4	
HIAL	DEV-HI	P5	
LOAL	DEV-LO	P6	
ROCHIAL	ROC-HI	P7	
ROCLOAL	ROC-LO	P8	
HIAL	HI-LIM	P9	
LOAL	LO-LIM	P10	

Figure K.3
General status word

If no consistency of bit positions for particular alarms is required, they may be used differently, for example according to priorities.

FREQTOTC

Converts a frequency from the frequency domain into the equivalent time constant in the time domain: TC = 0.1591549/FREQ

P1   Output TC of float type.
P2   Input FREQ of float type.

TCTOFREQ

Converts a time constant from the time domain into the equivalent frequency in the frequency domain: FREQ = 0.1591549/TC

P1   Output FREQ of float type.
P2   Input TC of float type.

EUTOPCT

Conversion from engineering units to %

P1   Output in %
P2   Input in engineering units.

P3   100% of range in engineering units.
P4   0% of range in engineering units.

PCTTOEU

Conversion from % to engineering units

P1   Output in engineering units.
P2   Input in %.
P3   100% of range in engineering units.
P4   0% of range in engineering units.

LINK

If the function (P3 XOR P3) = TRUE, the output P1 is set to the value of the input P2. If the function is NOT TRUE, P1 stays unchanged.

P1   Output of float type.
P2   Input of float type.
P3   Switch variable of status type.
P4   Constant, not related to any type variable:
P4 = 0   If switch P3 is TRUE, P1 will be set by P2
P4 = 1   If switch P3 is NOT TRUE, P1 will not be modified at all.

MATH

General purpose mathematical algorithm:

$$OP = (K1 \times X1 + K2 \times X2)/(K3 \times X3 + K4 \times X4) + K5 \times X5$$

Any unwanted part of the formula can be rendered ineffective by initialization of gain (K1, K2 etc.) to a value of 0.

P1   Output of float type.
P2   Input X1 of float type for 1st product.
P3   Gain K1 of float type for 1st product.
P4   Input X2 of float type for 2nd product.
P5   Gain K2 of float type for 2nd product.
P6   Input X3 of float type for 3rd product.
P7   Gain K3 of float type for 3rd product.
P8   Input X4 of float type for 4th product.
P9   Gain K4 of float type for 4th product.
P10   Input X5 of float type for 5th product.
P11   Gain K5 of float type for 5th product.

DIV

Divide algorithm: OP = NUMERATOR/DENOMINATOR

If the absolute value of the DENOMINATOR is less than 0.1% of the NUMERATOR, the OP value is set to BAD.

P1   OP of float type.
P2   NUMERATOR of float type.
P3   DENOMINATOR of float type.

HEATCOMP

Calculates the required FUEL flow for a feed heater to heat up feed material from T-IN (feed inlet temperature) to T-OUT (feed outlet temperature). Gain K represents a scaling factor, taking into account things such as fuel efficiency, etc. FEED is the feed flow, which may vary over time.

$$\text{FUEL} = K \times \text{FEED} \times (\text{T-OUT} - \text{T-IN})$$

P1   Output FUEL of float type.
P2   FEED of float type.
P3   T-OUT of float type.
P4   T-IN of float type.
P5 K of float type.

HEATSIM

Calculation for process simulation purposes only. It simulates the feed outlet temperature T-OUT of a feed heater (see HEATCOMP algorithm description above as well).

$$\text{T-OUT} = K \times \text{FUEL} / (\text{FEED} \times 1) + \text{T-IN}$$

P1   Output T-OUT of float type.
P2   FUEL of float type.
P3   FEED of float type.
P4   T-IN of float type.
P5 K of float type.

MASSFLOW

Massflow compensation, taking into account the volumetric flow (in real life: flow meter value), the temperature and pressure of the material. Gain K represents a general scaling factor, taking into account parameters such as specific mass, different engineering units, etc.

$$\text{MFLOW} = K \times \text{FLOW} \times (\text{PRESS}/(\text{TEMP} - 273.15))^{1/2}$$

P1   Output MFLOW of float type (massflow).
P2   FLOW of float type (volumetric flow).
P3   PRESS of float type (pressure).
P4   TEMP of float type (temperature).
P5   Gain K of float type.

DEADTIME

General purpose deadtime calculation as used in industrial controllers.

P1   Output of float type.
P2   Input of float type.
P3   Deadtime in minutes (float).

P4 Internal pointer variable. This variable itself is an integer type, but its contents is the number of the first of 22 sequential variables of float type. These 22 sequential variables are the deadtime history buffer. They have to be reserved for history collection and *must not* be used otherwise.

P5 Internal pointer variable. This variable itself is of integer type, but its content is the number of the float variable holding the oldest history value within the deadtime buffer.

Example:

```
*INT23:BUFFIDX        v = 132;   deadtime buffer: float 132 to 153 min
*FLOAT34:DEADTIME     v = 0.3;
*FLOAT35:INPUT        v = 50;
*FLOAT83:OUTPUT       v = 50;
*ALGO5:DEADTIME       P1 = 83   P2 = 35   P3 = 34   P4 = 23   P5 = 24;
```

The parameters P4 and P5 of the DEADTIME algorithm have to be of integer type. Their purpose is to provide pointers for calculating history data. Parameter P4 contains the index number of the first of 22 sequential floating point variables to be used as dead time history buffer. Parameter P5 contains the index number of the floating point variable containing the oldest history value. No parameter definition for integer 24 is required, since integer 24 is used as an internal parameter of DEADTIME and neither initialization nor display is necessary. There is no parameter definition required for float variables 132–153 as well.

K.2.1 General notes concerning algorithms

Multiple use of variables

The same variable may be used within one algorithm several times as different command parameters. This is only possible, if the parameters are of the same type. It may give different results as shown in the following two examples of the NOISE algorithm.

Example 1:

```
*FLOAT32:PV           v = 50;
*FLOAT33:PV-NOISE     v = 50;
*FLOAT34:K-NOISE      v = 1;
*ALGO2:NOISE          P1 = 33   P2 = 32   P3 = 34;
```

In this example, assuming that PV (float 32) or PV-NOISE (float 33) is not manipulated, PV-NOISE will stay within the range of 50 ± 0.5. Hence PV-NOISE will not drift away from 50.

Example 2:

```
*FLOAT32:PV           v = 50;
*FLOAT34:K-NOISE      v = 1;
*ALGO2:NOISE          P1 = 32   P2 = 32   P3 = 34;
```

In this example, assuming that PV (float 32), which is used as P1 and P2, is manipulated, then PV will drift away from 50 towards unpredictable values. A trend display of PV will show an unpredictable, non-cyclic trend.

Incremental algorithms

The incremental algorithms PIDN, PIDX, PROP, INT, DERN, DERX and ILIM may have the same variable (of float type) configured as P1 (output). If the same variable is jointly used as P1 by two or more of these incremental algorithms, the value of P1 represents the output of an incremental 'Summer' of the individual outputs. The following example shows this, using the algorithms PROP and INT to configure a simple PI-controller without alarm, limit and status handling.

Example:

*FLOAT41:SETPOINT	v = 50;			
*FLOAT42:PV	v = 50;			
*FLOAT43:REVERSE	v = − 1;	v = − 1	for reverse control action	
*FLOAT44:PROP-GAIN	v = 1.5;			
*FLOAT45:ERROR	v = 0;			
*FLOAT46:INTERNAL-USE	v = 50;	internal use of algorithm PROP		
*FLOAT47:INT-TC	v = 1.5			
*FLOAT49:OUTPUT	v = 50;			

*ALGO1:RATIO	P1 = 45	P2 = 42	P3 = 43	P4 = 42;
*ALGO2:PROP	P1 = 49	P2 = 45	P3 = 44	P4 = 46;
*ALGO3:INT	P1 = 49	P2 = 45	P3 = 47	

It is important to notice, that both algorithms, PROP and INT use the same variable as their output. The end result in the output variable OUTPUT is an output value which has been incremented (or decremented) by both algorithms. This is in actual fact an incremental 'Summer' of proportional and integral control actions.

Appendix L

Background graphics design

The design of background graphics requires the following steps:

- Use any paint program (e.g. Paint-Brush), which is capable of producing PCX-files. Create a PCX-file (filename.PCX) containing the desired graphics background. The graphics created has to have a screen resolution of 640 × 350 pixels in 16 color mode.
- Copy the file into the same directory in which the program file CONTROL.EXE is. If this step is omitted, the filename including path has to be configured.
- Configure display background, using the configuration command '*DISPLAY#. BACKGRND:filename;' within the PROCESS configuration file 'confname.PCF'.

Appendix M

Configuration example

M.1 Example file (FLOW.PCF)

M.1.1 Controller algorithms and variables

*ALGO1:LAG	P1 = 14	P2 = 40	P3 = 10;				
*ALGO2:PV	P1 = 14	P2 = 15	P3 = 4	P4 = 2	P5 = 3	P6 = 1	P7 = 8
	P8 = 1	P9 = 5	P10 = 6	P11 = 24	P12 = 25	P13 = 26	P14 = 27
	P15 = 28	P16 = 29;					
*ALGO3:PIDN	P1 = 11	P2 = 12	P3 = 14	P4 = 15	P5 = 7	P6 = 8	P7 = 9
	P8 = 2	P9 = 16	P10 = 17	P11 = 18;			
*ALGO4:OP	P1 = 1	P2 = 11	P3 = 12	P4 = 1	P5 = 3	P6 = 4	P7 = 1
	P8 = 2	P9 = 22	P10 = 23	P11 = 20	P12 = 21	P13 = 13;	
*ALGO5:ALARM	P1 = 6	P2 = 7	P3 = 1;				
*ALGO6:MODE	P1 = 5	P2 = 1	P3 = 1	P4 = 2	P5 = 8;		

*FLOAT1:OP	v = 50;	
*FLOAT2:PVE	v = 300;	
*FLOAT3:SPE	v = 300;	
*FLOAT4:CSP	v = 60;	
*FLOAT5:EUHI	v = 500;	
*FLOAT6:EULO	v = 0;	
*FLOAT7:K	v = 0.8;	
*FLOAT8:T_{INT}	v = 0.1;	
*FLOAT9:T_{DER}	v = 0;	
*FLOAT10:TD	v = 0.05;	
*FLOAT11:CVPD	v = 50;	
*FLOAT12:CVI	v = 0;	
*FLOAT13:OPVIRT	v = 50;	
*FLOAT14:PV	v = 60;	%
*FLOAT15:SP	v = 60;	%
*FLOAT16:LASTD	v = 0;	
*FLOAT17:LASTP	v = 0;	
*FLOAT18:LRATE	v = 0;	Used in PIDX algorithm only!
*FLOAT20:OPHI	v = 100;	
*FLOAT21:OPLO	v = 0;	
*FLOAT22:IHI	v = 95;	
*FLOAT23:ILO	v = 0;	
*FLOAT24:PVHI	v = 400;	

```
*FLOAT25:PVLO          v = 100;
*FLOAT26:DEVHI         v = 10;
*FLOAT27:DEVLO         v = − 10;
*FLOAT28:SPHI          v = 450;
*FLOAT29:SPLO          v = 50;
```

```
*STATUS1:MODE          v = 1            s0 = ÇMANUALÇ      s1 = ÇÇAUTOÇÇ
                       s2 = ÇÇCASCÇÇ
                       s3 = ÇI-MANÇÇ     s4 = ÇI-AUTOÇ      s5 = ÇI-CASCÇ;
*STATUS2:EQUATION      v = 0            s0 = ÇTYPEÇAÇ      s1 = ÇTYPEÇBÇ
                       s2 = ÇTYPEÇCÇ;
*STATUS3:ACTION        v = 1            s0 = ÇDIRECTÇ      s1 = ÇREVERSE;
*STATUS4:OPCALC.       v = 1            s0 = ÇÇREALÇÇ      s1 = ÇÇVIRTÇÇ;
*STATUS5:INIT          v = 0            s0 = ÇÇÇÇÇÇÇ       s1 = ÇÇINITÇÇ;
*STATUS6:PVALARM       v = 0            s0 = ÇÇÇÇÇÇÇÇ      s1 = INSTÇHIÇ
                       s2 = INSTÇLOÇ
                       Çs3 = ÇROCÇHIÇ    s4 = ÇROCÇLOÇ      s5 = ÇPVÇHIÇÇ
                       s6 = ÇPVÇLOÇÇ
                       s7 = ÇDEVÇHIÇ     s8 = ÇDEVÇLOÇ,
*STATUS7:OPALARM       v = 0            s0 = ÇÇÇÇÇÇÇ       s1 = ÇOPÇHIÇ
                       s2 = ÇOPÇLOÇÇ
                       s3 = ÇINTÇHIÇ     s4 = ÇINTÇLOÇ;
*STATUS8:CONFIG        v = 0            s0 = ÇÇÇÇÇÇÇÇ      s1 = ÇÇINITÇÇ
                       s2 = TRACKING
                       s3 = ÇIÇ&ÇTRÇ;
```

```
*INT1:STATUS           v = 0;
*INT2:2NDSTAT          v = 0;
```

Opening display

*DISPLAY0.BACKGRND:cover;

Tuning display

*DISPLAY5.BACKGRND:tuning;

Block diagram

```
*DISPLAY1.BACKGRND:flow;
*DISPLAY1.UPDATE:stv1      col = 19   row = 12   fg = HI-YELLOW   bg = BROWN;   MODE
*DISPLAY1.UPDATE:stdesc1   col = 15   row = 12   fg = HI-YELLOW   bg = BROWN;
*DISPLAY1.UPDATE:fv1       col = 27   row = 7    fg = HI-YELLOW   bg = BROWN;   OP
*DISPLAY1.UPDATE:fdesc1    col = 25   row = 7    fg = HI-YELLOW   bg = BROWN;
*DISPLAY1.UPDATE:fv2       col = 12   row = 15   fg = HI-CYAN     bg = CYAN;    PVE
*DISPLAY1.UPDATE:fdesc2    col = 9    row = 15   fg = HI-CYAN     bg = CYAN;
*DISPLAY1.UPDATE:fv3       col = 12   row = 3    fg = HI-GREEN    bg = GREEN;   SPE
*DISPLAY1.UPDATE:fdesc3    col = 9    row = 3    fg = HI-GREEN    bg = GREEN;
*DISPLAY1.UPDATE:stv6      col = 17   row = 5    fg = HI-RED      bg = BLACK;   PVALARM
```

M.1.2   Controller detail display

```
*DISPLAY2.BACKGRND:detail;
```

*DISPLAY2.DESC:Controller,	col = 1	row = 24	fg = HI-CYAN	bg = BLUE;	
*DISPLAY2.UPDATE:stv7	col = 6	row = 22	fg = HI-RED	bg = BLACK;	OPALARM
*DISPLAY2.UPDATE:stv1	col = 6 MODE	row = 21	fg = HI-YELLOW	bg = BLUE;	
*DISPLAY2.UPDATE:stdesc1	col = 1	row = 21	fg = HI-GREEN	bg = BLACK;	
*DISPLAY2.UPDATE:fv1	col = 6	row = 20	fg = HI-CYAN	bg = BLUE;	OP
*DISPLAY2.UPDATE:fdesc1h	col = 1	row = 20	fg = HI-YELLOW	bg = BLACK;	
*DISPLAY2.UPDATE:fv2	col = 6	row = 19	fg = HI-CYAN	bg = BLUE;	PVE
*DISPLAY2.UPDATE:fdesc2	col = 1	row = 19	fg = HI-CYAN	bg = BLACK;	
*DISPLAY2.UPDATE:fv3	col = 6	row = 18	fg = HI-CYAN	bg = BLUE;	SPE
*DISPLAY2.UPDATE:fdesc3	col = 1	row = 18	fg = HI-GREEN	bg = BLACK;	
*DISPLAY2.UPDATE:stv6	col = 6	row = 17	fg = HI-RED	bg = BLACK;	PVALARM
*DISPLAY2.UPDATE:fv7	col = 18	row = 5	fg = HI-CYAN	bg = BLUE;	K
*DISPLAY2.UPDATE:fdesc7	col = 14	row = 5	fg = HI-GREEN	bg = BLACK;	
*DISPLAY2.UPDATE:fv8	col = 18	row = 7	fg = HI-CYAN	bg = BLUE;	T_{INT}
*DISPLAY2.UPDATE:fdesc8	col = 14	row = 7	fg = HI-GREEN	bg = BLACK;	
*DISPLAY2.UPDATE:fv9	col = 18	row = 9	fg = HI-CYAN	bg = BLUE;	T_{DER}
*DISPLAY2.UPDATE:fdesc9	col = 14	row = 9	fg = HI-GREEN	bg = BLACK;	
*DISPLAY2.UPDATE:fv10	col = 18	row = 11	fg = HI-CYAN	bg = BLUE;	TD
*DISPLAY2.UPDATE:fdesc10	col = 14	row = 11	fg = HI-GREEN	bg = BLACK;	
*DISPLAY2.UPDATE:fv5	col = 33	row = 3	fg = HI-CYAN	bg = BLUE;	EUHI
*DISPLAY2.UPDATE:fdesc5	col = 28	row = 3	fg = HI-GREEN	bg = BLACK;	
*DISPLAY2.UPDATE:fv6	col = 33	row = 4	fg = HI-CYAN	bg = BLUE;	EULO
*DISPLAY2.UPDATE:fdesc6	col = 28	row = 4	fg = HI-GREEN	bg = BLACK;	
*DISPLAY2.UPDATE:fv24	col = 33	row = 6	fg = HI-CYAN	bg = BLUE;	PVHI
*DISPLAY2.UPDATE:fdesc24	col = 28	row = 6	fg = HI-GREEN	bg = BLACK;	
*DISPLAY2.UPDATE:fv25	col = 33	row = 7	fg = HI-CYAN	bg = BLUE;	PVLO
*DISPLAY2.UPDATE:fdesc25	col = 28	row = 7	fg = HI-GREEN	bg = BLACK;	
*DISPLAY2.UPDATE:fv26	col = 33	row = 8	fg = HI-CYAN	bg = BLUE;	DEVHI
*DISPLAY2.UPDATE:fdesc26	col = 28	row = 8	fg = HI-GREEN	bg = BLACK;	
*DISPLAY2.UPDATE:fv27	col = 33	row = 9	fg = HI-CYAN	bg = BLUE;	DEVLO
*DISPLAY2.UPDATE:fdesc27	col = 28	row = 9	fg = HI-GREEN	bg = BLACK;	
*DISPLAY2.UPDATE:fv28	col = 33	row = 11	fg = HI-CYAN	bg = BLUE;	SPHI
*DISPLAY2.UPDATE:fdesc28	col = 28	row = 11	fg = HI-GREEN	bg = BLACK;	
*DISPLAY2.UPDATE:fv29	col = 33	row = 12	fg = HI-CYAN	bg = BLUE;	SPLO
*DISPLAY2.UPDATE:fdesc29	col = 28	row = 12	fg = HI-GREEN	bg = BLACK;	
*DISPLAY2.UPDATE:fv22	col = 33	row = 14	fg = HI-CYAN	bg = BLUE;	IHI
*DISPLAY2.UPDATE:fdesc22	col = 28	row = 14	fg = HI-GREEN	bg = BLACK;	
*DISPLAY2.UPDATE:fv23	col = 33	row = 15	fg = HI-CYAN	bg = BLUE;	ILO
*DISPLAY2.UPDATE:fdesc23	col = 28	row = 15	fg = HI-GREEN	bg = BLACK;	
*DISPLAY2.UPDATE:fv20	col = 33	row = 16	fg = HI-CYAN	bg = BLUE;	OPHI
*DISPLAY2.UPDATE:fdesc20	col = 28	row = 16	fg = HI-GREEN	bg = BLACK;	
*DISPLAY2.UPDATE:fv21	col = 33	row = 17	fg = HI-CYAN	bg = BLUE;	OPLO
*DISPLAY2.UPDATE:fdesc21	col = 28	row = 17	fg = HI-GREEN	bg = BLACK;	
*DISPLAY2.UPDATE:stv2	col = 29	row = 23	fg = HI-YELLOW	bg = BLACK;	EQUATION
*DISPLAY2.UPDATE:stdesc2	col = 21	row = 23	fg = HI-YELLOW	bg = BLACK;	
*DISPLAY2.BAR:2	col = 11		fg = HI-YELLOW	bg = BLACK;	
*DISPLAY2.BAR:1	col = 9		fg = HI-CYAN	bg = BLACK;	
*DISPLAY2.MARKER:0	col = 8		fg = HI-GREEN	bg = BLACK;	

M.1.3 Auxiliary screen

*DISPLAY7.BACKGRND:Auxiliary;

Left column for control variables

*DISPLAY7.UPDATE:fv11	col = 10	row = 2	fg = HI-CYAN	bg = BLUE;	CVPD
*DISPLAY7.UPDATE:fdesc11	col = 2	row = 2	fg = HI-GREEN	bg = BLUE;	
*DISPLAY7.UPDATE:fv12	col = 10	row = 3	fg = HI-CYAN	bg = BLUE;	CVI
*DISPLAY7.UPDATE:fdesc12	col = 2	row = 3	fg = HI-GREEN	bg = BLUE;	
*DISPLAY7.UPDATE:fv13	col = 10	row = 4	fg = HI-CYAN	bg = BROWN;	OPVIRT
*DISPLAY7.UPDATE:fdesc13	col = 2	row = 4	fg = HI-GREEN	bg = BROWN;	
*DISPLAY7.UPDATE:fv14	col = 10	row = 5	fg = HI-CYAN	bg = BLUE;	PV
*DISPLAY7.UPDATE:fdesc14	col = 2	row = 5	fg = HI-GREEN	bg = BLUE;	
*DISPLAY7.UPDATE:fv15	col = 10	row = 6	fg = HI-CYAN	bg = BLUE;	SP
*DISPLAY7.UPDATE:fdesc15	col = 2	row = 6	fg = HI-GREEN	bg = BLUE;	
*DISPLAY7.UPDATE:fv16	col = 10	row = 7	fg = HI-CYAN	bg = BLUE;	LASTD
*DISPLAY7.UPDATE:fdesc16	col = 2	row = 7	fg = HI-GREEN	bg = BLUE;	
*DISPLAY7.UPDATE:fv17	col = 10	row = 8	fg = HI-CYAN	bg = BLUE;	LASTP
*DISPLAY7.UPDATE:fdesc17	col = 2	row = 8	fg = HI-GREEN	bg = BLUE;	
*DISPLAY7.UPDATE:stv1	col = 10	row = 10	fg = HI-YELLOW	bg = BLUE;	MODE
*DISPLAY7.UPDATE:stdesc1	col = 2	row = 10	fg = HI-GREEN	bg = BLUE;	
*DISPLAY7.UPDATE:stv2	col = 10 EQUATION	row = 11	fg = HI-MAGENTA	bg = BLUE;	
*DISPLAY7.UPDATE:stdesc2	col = 2	row = 11	fg = HI-GREEN	bg = BLUE;	
*DISPLAY7.UPDATE:stv3	col = 10 ACTION	row = 12	fg = HI-MAGENTA	bg = BLUE;	
*DISPLAY7.UPDATE:stdesc3	col = 2	row = 12	fg = HI-GREEN	bg = BLUE;	
*DISPLAY7.UPDATE:stv4	col = 10	row = 13	fg = HI-MAGENTA	bg = BLUE;	OPLIMIT
*DISPLAY7.UPDATE:stdesc4	col = 2	row = 13	fg = HI-GREEN	bg = BLUE;	
*DISPLAY7.UPDATE:stv5	col = 10	row = 14	fg = HI-CYAN	bg = BLUE;	INIT
*DISPLAY7.UPDATE:stdesc5	col = 2	row = 14	fg = HI-GREEN	bg = BLUE;	
*DISPLAY7.UPDATE:stv6	col = 10	row = 15	fg = HI-RED	bg = BLUE;	PVALARM
*DISPLAY7.UPDATE:stdesc6	col = 2	row = 15	fg = HI-GREEN	bg = BLUE;	
*DISPLAY7.UPDATE:stv7	col = 10	row = 16	fg = HI-RED	bg = BLUE;	OPALARM
*DISPLAY7.UPDATE:stdesc7	col = 2	row = 16	fg = HI-GREEN	bg = BLUE;	
*DISPLAY7.UPDATE:stv8	col = 10	row = 17	fg = HI-MAGENTA	bg = BLUE;	CONFIG
*DISPLAY7.UPDATE:stdesc8	col = 2	row = 17	fg = HI-GREEN	bg = BLUE;	

Right column for simulation variables

*DISPLAY7.UPDATE:fv31	col = 32	row = 2	fg = HI-CYAN	bg = BLUE;	LAG1TC
*DISPLAY7.UPDATE:fdesc31	col = 24	row = 2	fg = HI-GREEN	bg = BLUE;	
*DISPLAY7.UPDATE:fv34	col = 32	row = 3	fg = HI-CYAN	bg = BLUE;	LAG1VAL
*DISPLAY7.UPDATE:fdesc34	col = 24	row = 3	fg = HI-GREEN	bg = BLUE;	
*DISPLAY7.UPDATE:fv39	col = 32	row = 14	fg = HI-MAGENTA	bg = BLUE;	K-NOISE
*DISPLAY7.UPDATE:fdesc39	col = 24	row = 14	fg = HI-GREEN	bg = BLUE;	
*DISPLAY7.UPDATE:fv40	col = 32	row = 17	fg = HI-CYAN	bg = BLUE;	SIM-VAL
*DISPLAY7.UPDATE:fdesc40	col = 24	row = 17	fg = HI-GREEN	bg = BLUE;	

M.1.4   Trend display

*DISPLAY3.BACKGRND:trend;

*DISPLAY3.DESC:Trend Display,	col = 1	row = 24	fg = HI-CYAN
	bg = BL;		

Right side

*DISPLAY3.UPDATE:fv3	col = 33	row = 17	fg = HI-CYAN	bg = BLUE;	SP
*DISPLAY3.UPDATE:fdesc3	col = 28	row = 17	fg = GREEN	bg = BLACK;	
*HISTORY0:fv3	tmin = 0	tmax = 500	c = GREEN;		
*DISPLAY3.UPDATE:stv6	col = 19	row = 18	fg = HI-RED	bg = BLACK;	PVALARM
*DISPLAY3.UPDATE:fv2	col = 33	row = 18	fg = HI-CYAN	bg = BLUE;	PV
*DISPLAY3.UPDATE:fdesc2	col = 28	row = 18	fg = HI-GREEN	bg = BLACK;	
*HISTORY1:fv2	tmin = 0	tmax = 500	c = HI-GREEN;		
*DISPLAY3.UPDATE:stv7	col = 19	row = 19	fg = HI-RED	bg = BLACK;	OPALARM
*DISPLAY3.UPDATE:fv1	col = 33	row = 19	fg = HI-YELLOW	bg = BLUE;	OP
*DISPLAY3.UPDATE:fdesc1	col = 28	row = 19	fg = HI-YELLOW	bg = BLACK;	
*HISTORY2:fv1	tmin = 0	tmax = 100	c = HI-YELLOW;		

Left side

*DISPLAY3.UPDATE:stv1	col = 6	row = 19	fg = HY	bg = BLUE;
	MODE			
*DISPLAY3.UPDATE:stdesc1	col = 1	row = 19	fg = HY	bg = BLUE;

M.1.5   Process simulation

*ALGO10:LAG	P1 = 34	P2 = 1	P3 = 31;		
	First Lag as Dynamic Sim				
*ALGO16:NOISE	P1 = 34	P2 = 34	P3 = 39;		
	Incremental random as noise				
*ALGO17:PCTTOEU	P1 = 40	P2 = 34	P3 = 37	P4 = 38;	P1 = 40 is
	Output of Sim in PCT				

*FLOAT31:LAG1TC	v = 0.1;
*FLOAT34:LAG1VAL	v = 50;
*FLOAT37:SIMEUHI	v = 120;
*FLOAT38:SIMEULO	v = 0;
*FLOAT39:K-NOISE	v = 1;
*FLOAT40:SIM-VAL	v = 60;

Introduction to exercises

Starting training applications from DOS

Make the working directory the same as the one where the process control software has been installed. To do so, use the DOS command for changing directory (CD directory name). Type 'MENU' and a menu, showing available training applications will appear. Then, use the cursor keys on your keyboard to move the cursor to the training application you want to select. Press Enter and the exercise which has been selected, will come up.

Starting training applications from MS-Windows

Open the folder called CONTROL. Select and open the training application you want to run.

Gain of simulated disturbance

All training applications are configured with correct tuning constants and are ready for operation. The process simulation of each training application is configured with process disturbance process noise. The gain for disturbance is originally set to 0 and may be changed by the student in order to add more realism into the process simulations. A value of 1 is recommended for 'K-DIST' (gain of process disturbance).

The magnitude of disturbances can be changed to challenge the operational skills of the operating student. The disturbances will create realistic situations, where operator interventions on outputs and limits are necessary to avoid loss of control. In addition, the value of the process disturbance 'DISTURB' can be changed by the student directly. This permits the simulation of step functions of the process disturbance.

Example

In a level control system, a flow controller controls the inlet flow into a container. The outlet flow is unpredictable and represents the process disturbance. If the outlet flow reaches 0 and a controller output low limit (OPLO) of 5% exists, a windup situation exists. The container will overflow because OP is limited to 5% and no outlet flow exists. Operator intervention is required.

For most exercises, the disturbances have to be left at 0. Call up auxiliary display F8 for any modification to the disturbance. If only one disturbance exists in the process simulation, the disturbance gain is called K-DIST. However, if more than one disturbance is simulated, then for each disturbance exists a separate gain (K-DIST-F, K-DIST-T, K-DIST-P, etc. for disturbance gain of *flow*, *temperature*, *pressure*, etc.). Reducing the gain of the disturbance makes operation easier and less of a challenge. In order to see the principles of control clearly, no process disturbance is desired. Therefore, the exercises described below require the student to leave gain for random disturbances at 0 (default at start-up).

Note: Stabilize the process simulation and controller before each exercise if necessary.

First, make sure disturbance gain is at 0 (display F8). To stabilize the process, change the controller MODE to MANUAL and set the OP to 50% (in cascade control, MODE and OP of flow controller only). Make sure the process is steady before you start with a particular exercise.

As it is of great importance to understand the control concepts employed in different applications, the exercises will make use of the concepts explained in the previous chapters. A selection of different applications will be used to demonstrate the use of those concepts.

Trend displays

The trend displays shown in the document are black and white only. This makes it very hard to interpret the meaning of multiple trend pens. The trend displays shown in the documentation have to be used together with the actual displays of the simulation.

The explanations given in the documentation are more useful when the student observes the colored trend display and the computer screen at the same time.

Exercise 1

Flow control loop – basic example

E.1.1   Objective

This exercise will familiarize the student with the basic concept of closed loop control. It provides an opportunity to get a first feel for closed loop control. A flow control loop will serve as a practical example for this exercise. A flow control loop is generally not critical to operate and illustrates the basic principles effectively.

E.1.2   Operation

Since this is a relatively simple exercise, it can be used for familiarization with the principal operation of the simulation software. Call up the training application *single flow loop* as explained above.

Now, press the F2 button to call up the flow control block diagram display. The display as shown in Figure Ex.1.1 will appear. This display gives a general idea of the process and displays all major variables.

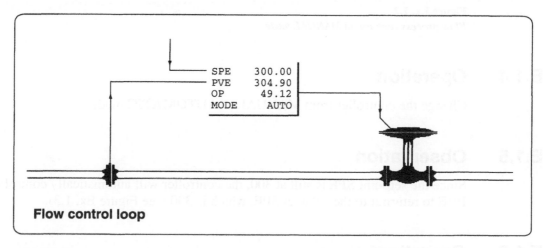

Flow control loop

SPE	300.00
PVE	304.90
OP	49.12
MODE	AUTO

Figure Ex. 1.1
Flow control display

First, we will observe the general behavior of the process. This is best done on a trend display. Call up the trend display of this exercise by pressing F4. At this stage the flow control is up and running correctly in automatic control. In order to observe the process reaction, as a result of changes in the position of the control valve, change the control mode from AUTO to MANUAL. Then, change the OP value of the controller to 20% as follows: Type the parameter name 'OP' followed by Enter. Then enter the value 15 representing 15% of output.

E.1.3   Observation

We observe the process variable PVE (process variable in engineering units) as it changes in value due to the change in OP. The PV will change as shown in Figure Ex. 1.2. The setpoint is not used in MANUAL mode and is ignored. The setpoint SPE (setpoint in engineering units) just happens to be 300 at this time.

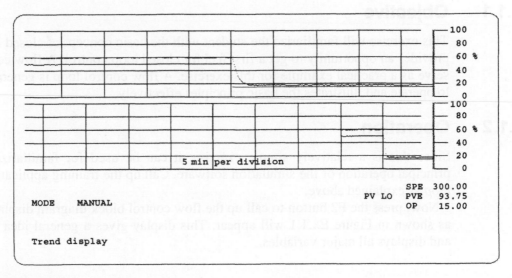

Figure Ex. 1.2
Flow process reaction in MANUAL mode

E.1.4   Operation

Change the controller from MANUAL to AUTOMATIC mode.

E.1.5   Observation

Since the setpoint SPE is still at 300, the controller will automatically control the value of PVE to return it to the value of SPE, which is 300 (see Figure Ex. 1.3).

E.1.6   Operation

A detail display can be called up with the F3 button (see Figure Ex. 1.4). The student should now experiment on his/her own. You are encouraged to make changes to range, limits, modes and values in order to experience their effect. Since a flow control loop has no intrinsic stability problems, most effects can be observed clearly.

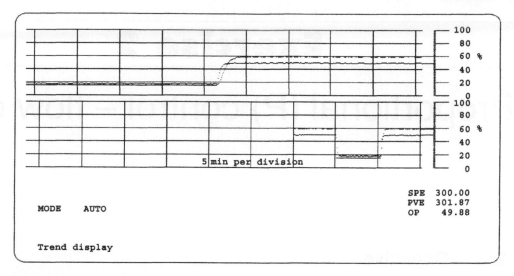

Figure Ex. 1.3
Flow control – automatic return off PVE to SPE

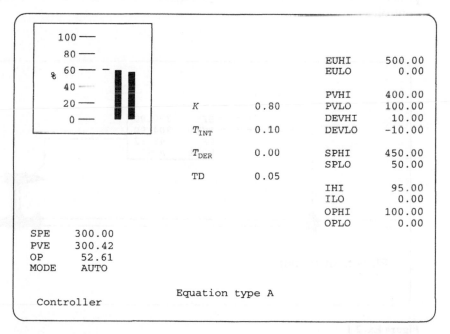

Figure Ex. 1.4
Flow controller detail display

The student may call up the auxiliary display F8 to look at some values that are important to the simulation. The auxiliary display shows two columns of variables; the left column shows important background variables of the controller and the right column shows simulation variables.

Exercise 2

Proportional (P) control – flow chart

E.2.1   Objective

This exercise will introduce the main control action of controllers – proportional control. Special emphasis is placed on the fact that there is a remaining offset condition, if proportional control is used solely. Figure Ex. 2.1 shows an example for closed loop control.

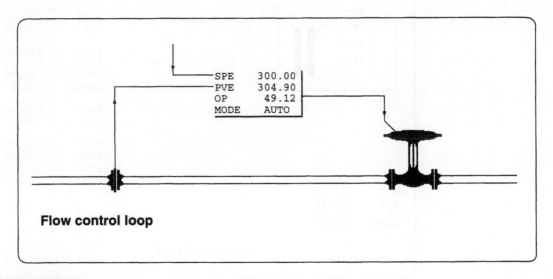

```
SPE     300.00
PVE     304.90
OP       49.12
MODE     AUTO
```

Flow control loop

Figure Ex. 2.1
Flow control loop

E.2.2   Operation

Call up the training application *single flow loop*. After this exercise has been called up, press F3 to get the detail display of the flow controller.

To prepare the controller for P-control only, change T_{INT} (integral time constant) and T_{DER} (derivative time constant). Set T_{INT} to almost infinity. Set T_{DER} to zero for no derivative action.

Practical values are 999 for T_{INT} and 0 for T_{DER}, as shown in Figure Ex. 2.2.

To study P-control, change SPE from 300 (60% of range) to 125 (25% of range).

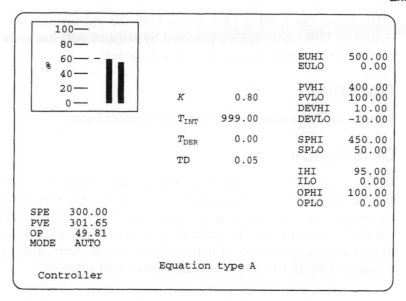

Figure Ex. 2.2
Flow controller detail display for P-control

E.2.3   Observation

It is observed, that the PVE value settles down before reaching the value of *SPE*. The remaining difference between SPE and PVE is called proportional offset. It can also be observed that the control action moves almost instantaneously (proportional control) in accordance with changes to PVE or SPE (equation type A). The trend display has to be observed very carefully (see Figure Ex. 2.3). The step change of SPE has caused the value of OP to make a step change accordingly. Immediately after this, one can observe the output to exactly follow the value of PVE, but in reverse direction to the PVE.

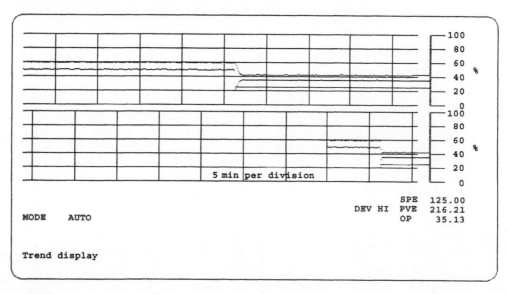

Figure Ex. 2.3
Proportional control trend with offset

Therefore, the trend display shows both the proportional change of output based on change of SPE (step change), followed by proportional change of output based on change of PVE (exponential approach to a steady value).

E.2.4   Operation

Repeat the above exercise with different values of gain.

E.2.5   Conclusion

With increasing values of gain, we obtain smaller values for OFFSET. This shows, that it may be desirable to use high values for gain to minimize the offset. If a control loop has a tendency to be unstable, stability problems put limits on the increase of gain. Even if no stability problem exists, the value of gain should be kept as low as possible. This avoids unnecessary amplification of noise. As a result, we learn that it is practically impossible to reduce OFFSET to zero in an industrial control situation.

Exercise 3

Integral (I) Control – flow control

E.3.1   Objective

This exercise will introduce the integral control action of controllers. Special emphasis is given to the task of eliminating the remaining offset term of proportional control. It will also show the slower control action of integral control, compared with proportional control. Figure Ex. 2.1 (see Exercise 2) shows an example for closed loop control.

E.3.2   Operation

Call up the training application *single flow loop*. After this exercise has been called up, press F3 to get the detail display of the flow controller.

To prepare the controller for I-control only, set K to 0, T_{INT} to 1 and T_{DER} to 0. $K = 0$ causes the controller to switch to integral control only, using a unit gain of 1 for integral only. Gain for proportional and derivative control action is 0, as the detail display shows (see Figure Ex. 3.1).

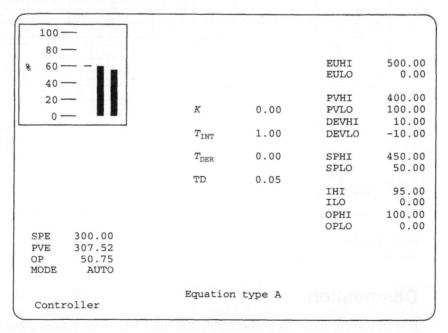

Figure Ex. 3.1
Flow controller detail display for I-control

To study I-control, change SPE from 300 (60% of range) to 125 (25% of range) as shown in Figure Ex. 3.2. Then change SPE back to 300 (A value of 310 has been used to create Figure Ex. 3.3).

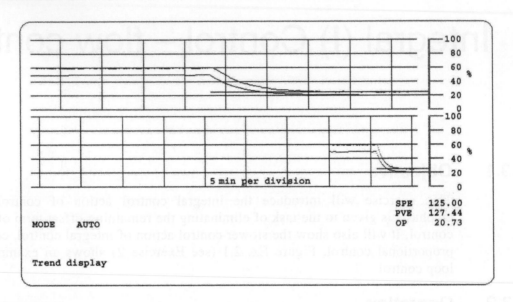

Figure Ex. 3.2
Integral control trend (SPE 300 to 150)

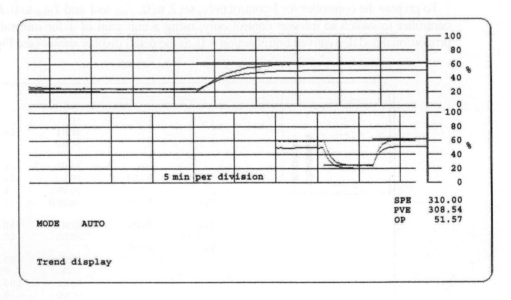

Figure Ex. 3.3
Integral control trend (SPE 150 to 310)

E.3.3   Observation

It is observed, with integral control, that the PVE value settles down exactly at the value of SPE. *No offset* is left, as would have been with proportional control only. It can also be observed that the control action does not start with a step, as a result of a step of SPE, as it did with proportional control. The control action resulting from integral control action is slow because it has a lagging behavior as shown in Figures Ex. 3.2 and 3.3.

E.3.4   Operation

Repeat the above exercise with different values of T_{INT}.

E.3.5   Conclusion

Integral control will control the PVE towards the SPE precisely, without any OFFSET. The trade-off for this, when compared with proportional control, is a lagging behavior, which results in slower control. It will be shown later, that integral control decreases the stability of the loop, if an intrinsic stability problem exists within the control loop. Integral control and its effects on stability are dealt with in Chapter 5, 'Tuning of closed loop control' and Exercises 6, 7 and 8.

Exercise 4

Proportional and integral (PI) control – flow control

E.4.1   Objective

This exercise will introduce a combined proportional and integral control action of controllers. Special emphasis is given to the elimination of the remaining offset of proportional control without loss of control speed. It will be shown that the combination of proportional and integral control maintains the speed of control as it exists with proportional control only, but without the disadvantage of an OFFSET term. Figure Ex. 2.1 shows an example for closed loop control.

E.4.2   Operation

Call up the training application *single flow loop*. After this exercise has been called up, press F3 to get the detail display of the flow controller (Figure Ex. 4.1).

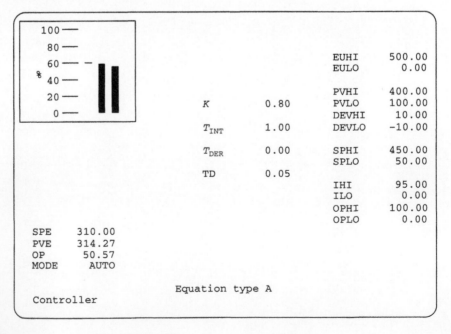

Figure Ex. 4.1
Flow controller detail display for PI-control

To prepare the controller for PI-control, change K (gain), T_{INT} (integral time constant) and T_{DER} (derivative time constant). Set K to 0.8, T_{INT} to 1 and T_{DER} to 0.

To study PI-control, change SPE from approximately 300 (60% of range) to approximately 125 (25% of range) as shown in Figure Ex. 4.2.

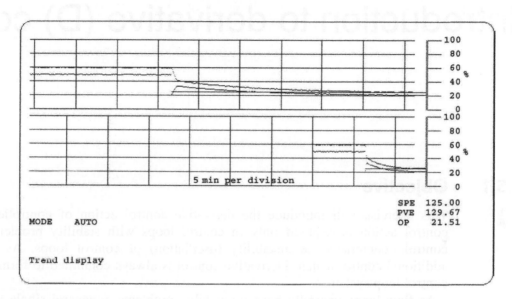

Figure Ex. 4.2
Combined proportional and integral control

E.4.3 Observation

It can be seen, that combined proportional and integral control moves the PVE value closely towards the value of SPE. *No offset* is left, as would have been with proportional control only. There has also been no noticeable loss of speed of control. Close observation of the trend display (see Figure Ex. 4.2) first reveals a fast change of PVE, followed by a slow approach towards SPE. The fast change of PVE is the direct result of proportional control and the slower change of PVE which follows is mainly the result of integral control.

E.4.4 Operation

Repeat the above exercise with different values of T_{INT}. Keep K below 1 and use small values for T_{INT} (in our example 0.05–0.5), as is done in most practical applications of flow loops.

E.4.5 Conclusion

Combined proportional and integral control turns out to be the best choice of control, if no stability problem exists within the control loop. As a general rule (with few exceptions), flow control loops have no stability problems. This is the reason why the majority of flow control loops use PI-control. Stability problems are dealt with in Chapter 5, 'Tuning of closed loop control systems' and Exercises 6, 7 and 8.

Exercise 5

Introduction to derivative (D) control

E.5.1   Objective

This exercise will introduce the derivative control action of controllers. Derivative control action is required only in control loops with stability problems. Derivative control counteracts the instability (oscillation) of control loops. As such it is an additional control action. Derivative control is always combined to form a PD or PID-control loop system.

As flow loops generally have no stability problems, a general single loop is used to demonstrate D-control. Figure Ex. 5.1 shows a single loop to control temperature. This is an example for a control loop, where a stability problem can be expected. It would create unrealistic responses if we used derivative control in a flow control loop. A general control loop is shown in Figure Ex. 5.2.

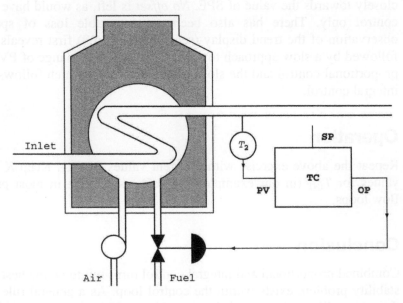

Figure Ex. 5.1
Feed heater single loop control

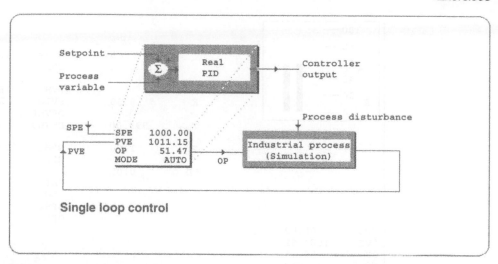

Figure Ex. 5.2
Single loop control block diagram

E.5.2 Operation

Call up the training application general single loop with interactive PID (real form).

To confirm K-DIST is at 0, call up display F8 (auxiliary display). If necessary, change the gain for the disturbance to 0 (K-DIST = 0). It is important to be aware that some process noise still exists (K NOISE = 0.1) (see Figure Ex. 5.3).

```
OPVIRT        49.79        LAG1TC          0.40
OP            49.79        LAG1VAL        49.80
PV            49.61        LAG2TC          0.40
CSP           50.00        LAG2VAL        49.79
MODE          AUTO
ACTION        REVERSE      K-DIST          0.00
OPCALC.       VIRT         TC-DIST         3.00
INIT                       DIST-HI        20.00
CONFIG                     DIST-LO       -20.00
                           DISTURB         0.00
                           RAW-DIST        0.00

                           SUM-DIST       49.79

                           LAG3TC          0.40
                           LAG3VAL        49.78
CVPD           0.02
CVI           -0.00        K-NOISE         0.10
                           SIM-VAL        49.58

      Control values              Simulation values
```

Figure Ex. 5.3
Auxiliary display (F8)

To study D-control alone, call up the detail display F3 and change K, T_{INT} and T_{DER} to the values shown in Figure Ex. 5.4.

In order to observe D-control, change SPE from approximately 1000 (50% of range) to approximately 1200 (60% of range). Figure Ex. 5.5 shows an example where SPE has been changed from 1000 to 1220.

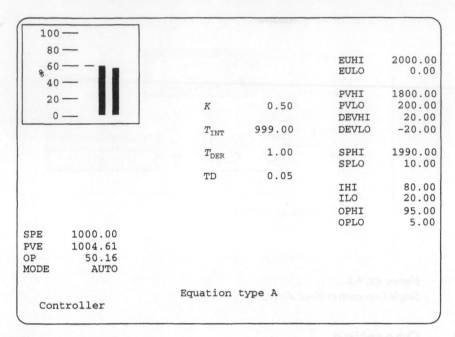

Figure Ex. 5.4
Controller detail display for D-control

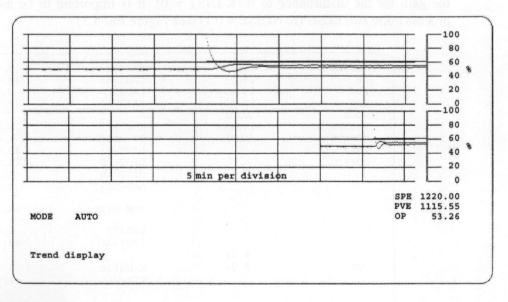

Figure Ex. 5.5
Trend display for D-control

E.5.3   Observation

It is observed that derivative control calculates a controller OP value based on the rate of change of PVE – SPE. As is explained in greater detail in the Chapter 'Digital control principles', D-control of an 'Interactive PID controller' is in actual fact a PD-control action with a low-pass filter (see Figure 6.5). Therefore, the initial OP change as a result of the step change of SPE is not a unit-pulse-function (needle shaped pulse with magnitude); it is in fact a limited step of OP, followed by an exponential decay. The initial step is $8 \times K \times (\text{PVE} - \text{SPE})$. As a P-component always exists within this control action, PVE settles down with an OFFSET. Exercise 10 will demonstrate the different kinds of D-control in more detail.

E.5.4 Operation

Repeat the above exercise with different values of T_{DER} and K. Observe the clipping of OP at the OPHI and OPLO limits, if large values of K are used.

E.5.5 Conclusion

Derivative control is based on the rate of change of PVE – SPE and is not designed to bring the value of PVE to the value of SPE. The sole purpose of D-control is to stabilize an intrinsically unstable control loop.

Exercise 6

Practical introduction into stability aspects

E.6.1 Objective

The objective of this exercise is to demonstrate the direct relationship between process lag and stability, and to make students aware of the great difference between noise and stability in practice.

E.6.2 Operation

Call up the training application General single loop with interactive PID (real form).

In order to see the stability problem of closed loops in practice and without disturbances, confirm that K-DIST is 0. Call up auxiliary display F8 and change the gain for disturbance to 0 (K-DIST = 0) and the gain for noise to 0 as well (K-NOISE = 0). Then the auxiliary display should be as shown in Figure Ex. 6.1.

```
OPVIRT        49.64          LAG1TC         0.40
OP            49.64          LAG1VAL       49.96
PV            49.82          LAG2TC         0.40
CSP           50.00          LAG2VAL       49.99
MODE          AUTO
ACTION        REVERSE        K-DIST         0.00
OPCALC.       VIRT           RTRN-TC       10.00
INIT                         DISTURB        0.00
CONFIG                       SUM-DIST      49.99

                             K-NOISE        0.00
                             LAG3TC         0.40
                             LAG3VAL       49.92

                             SIM-VAL       49.84


CVPD           0.02
CVI           -0.00


     Control values            Simulation values
```

Figure Ex. 6.1
Auxiliary Display Showing K-DIST = 0 and K-NOISE = 0

Call up detail display F3 as well, and make the changes $K = 3.5$, $T_{INT} = 999$ and $T_{DER} = 0$ as shown in Figure Ex. 6.2. The values shown in the detail display make it possible to create a continuous oscillation, purely based on the lag of the process. With these values, no integral and no derivative control action takes place and the controller will neither increase nor decrease the phase shift within the control loop. Therefore, the existing phase shift has been caused by the process only.

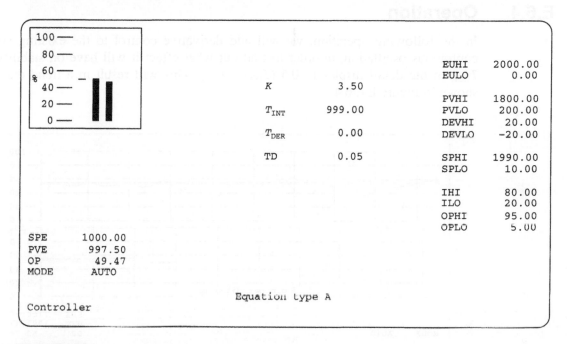

Figure Ex. 6.2
Detail display

Change *SPE* from 1000 to 1100 to start continuous oscillation (see Figure Ex. 6.3).

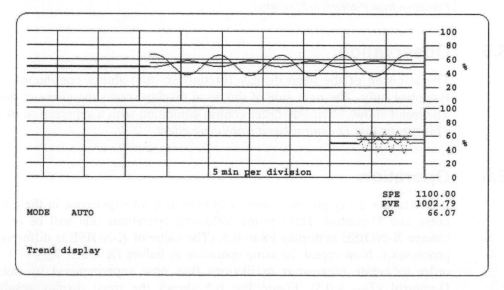

Figure Ex. 6.3
Continuous oscillation

E.6.3   Observation

In this situation, the controller performs P-control only. As shown in the Chapter 'Stability and control modes of closed loops', continuous oscillation exists if loop gain is 1 and loop phase shift is 180. Since our controller gain K is 3.5 we can calculate a process gain of 0.3.

E.6.4   Operation

In the following operation, we will add derivative control to the existing situation of continuous oscillation, in order to find out what effect it will have on stability. Change T_{DER} in the detail display to 0.5 (T_{DER} = 0.5). This will result in a trend display F4 as shown in Figure Ex. 6.4.

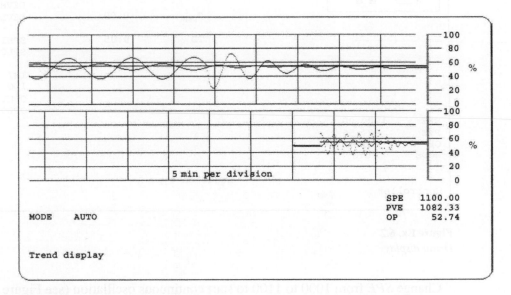

Figure Ex. 6.4
Transition from P-control to PD-control

E.6.5   Observation

The left side of the trend display (Figure Ex. 6.4) shows continuous oscillation with P-control only. The right side of the trend display shows the effect of added D-control. The trend shows obviously faster control combined with a suppression of the oscillation. This means that *D-control has a stabilizing effect.*

E.6.6   Operation

It is vital for every process control engineer to have experience in the effects of process noise and D-control. Through the following operations this will be experienced. First change K-NOISE in display F8 to 0.5. (The value of K-NOISE is different with different processes.) Now repeat the same operation as before (K = 3.5, T_{DER} + 0, T_{INT} + 999), in order to create continuous oscillations (but now superimposed by noise). Then add D-control (T_{DER} + 0.5). Figure Ex. 6.5 shows the trend display resulting from this exercise.

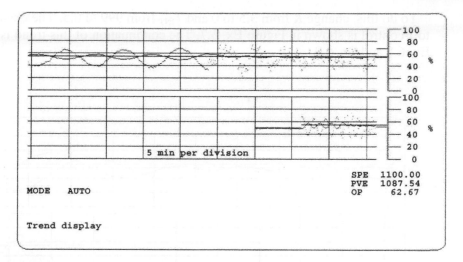

Figure Ex. 6.5
Transition from P-control to PD-control of a noisy process

E.6.7 Observation

It is clearly visible, that D-control multiplies noise.

E.6.8 Operation

Now repeat the exercise as above, but add I-control (not D-control as before), in a continuous oscillating situation created by P-control. The student adds I-control in order to change from P-control to PI-control. Use $T_{INT} = 0.5$ to obtain a trend display as shown in Figure Ex. 6.6. Then start again with continuous oscillation, created by P-control only and change to I-control only.

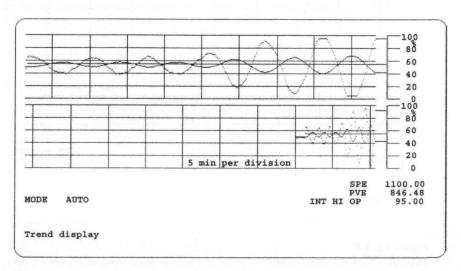

Figure Ex. 6.6
Transition from P-control to PI-control of a noisy process

To do this, change *K* from 3.5 to 0 and T_{INT} from 999 to 0.5. The change from P-control to I-control is shown in Figure Ex. 6.7. The continuation of this trend is shown in Figures Ex. 6.8 and Ex. 6.9.

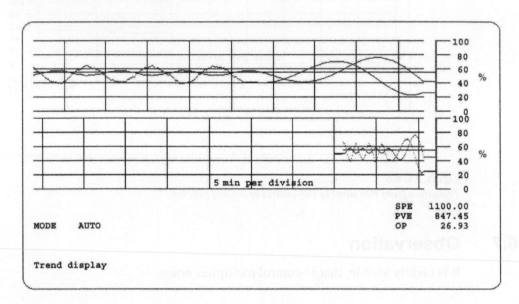

Figure Ex. 6.7
Transition from P-control to I-only control of a noisy process

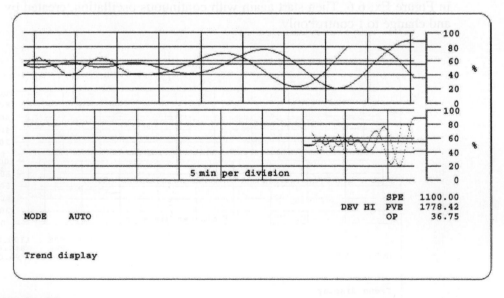

Figure Ex. 6.8
Transition from P-control to I-only control of a noisy process (cont. of Figure Ex. 6.7)

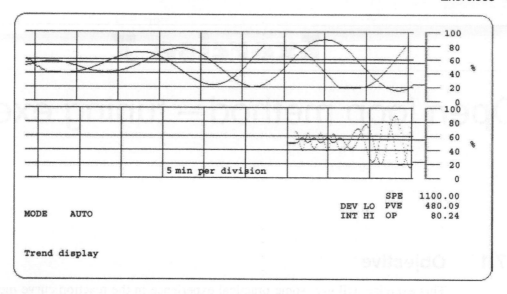

Figure Ex. 6.9
Transition from P-control to I-only control of a noisy process (cont. of Figure Ex. 6.8)

E.6.9   Observation

Figures Ex. 6.6–6.9 show that integral action has a destabilizing effect, whether it is combined PI-control or I-control only. In both cases we observe an increase of magnitude in oscillation. In addition we observe, that I-control ignores noise. We can see in Figure Ex. 6.6, that the change from P-control to PI-control has no effect on noise. Noise is not different in P-control and PI-control. Noise is almost totally suppressed when I-control is used only (see Figures Ex. 6.7–Ex. 6.9).

E.6.10   Conclusion

This exercise brings us to the following conclusions:

- D-control has a stabilizing effect, whereas I-control has a destabilizing effect. The only purpose of I-control is to eliminate the offset of P-only control.
- The trade-off is the destabilizing effect. This may be compensated by adding D-control. This will result in the use of PID-control for control loops with stability problems.
- D-control has to be treated very carefully from a noise point of view. It is important to point out that D-control should never be used without prior filtering of the PVE. Such a filter has to reduce noise without adding a significant Lag to the loop. The student will find in the detail display F3 a variable TD. TD is the filter time constant for PV filtering.

Note: Never use D-control without PV filtering (TD) if any process noise exists.

Exercise 7

Open loop method – tuning exercise

E.7.1 Objective

This exercise will give some practical experience in the reaction curve method of tuning. Figure Ex. 7.1 shows the diagram of a typical control loop. The steps for tuning are explained in Chapter 5, 'Tuning of closed loop control systems'.

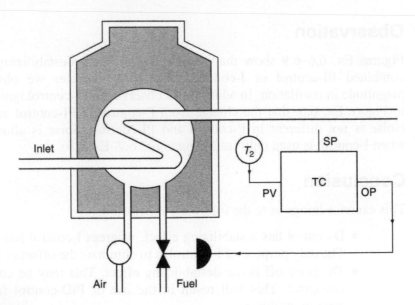

Figure Ex. 7.1
Single loop control

E.7.2 Operation

Call up the training application *General single loop with interactive PID (real form)*.

Call up the detail display F3 and change MODE to MANUAL. Make a good judgment about the noise observed on the PV and set the digital noise filter (TD) accordingly. A good value of TD will result in a still noticeable but very small magnitude of noise on PV. The noise filter has to be set up before tuning in order to have it included in the result of tuning. From a practical point of view, the noise filter becomes part of the process as far as loop tuning is concerned. In this exercise, a good value for TD will be between 0.05 and 0.1.

Make a small change of OP, to find out how sensitive the process is in its response. Then wait until the process is steady again (until PV is fairly constant). Now make a change of OP with a magnitude which ensures that the PV stays within its alarm limits PVHI and PVLO (Ignore the violation of *deviation* limits in this exercise).

To obtain a process reaction curve as shown in Figure Ex. 7.2, set MODE to MANUAL, then OP to 50 and wait until the PVE is steady at approximately 1000 (50% of range). Then make a change of the OP from 50 to 65.

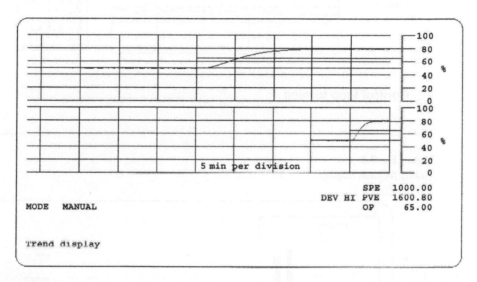

Figure Ex. 7.2
Reaction curve trend display for open loop tuning

E.7.3   Observation

Observe the reaction curve and tune the controller as explained in Chapter 5, 'Tuning of closed loop control'. Call up display F6 to get the summary of the major tuning formulas on screen as shown in Figure Ex. 7.3.

The values obtained from the reaction curve are:

$$\Delta OP = 15\% \text{ (from 50 to 65\%)}$$
$$N = 20\%/\text{min (PVE range from 0 to 2000)}$$
$$L = 0.4 \text{ min}$$

Using the formulas shown in display F6, we obtain the following tuning constants for PID-control:

$$K \quad = 2.2$$
$$T_{INT} = 0.8$$
$$T_{DER} = 0.2$$

After setting the tuning constants to their calculated values, the detail display F3 should look as shown in Figure Ex. 7.4.

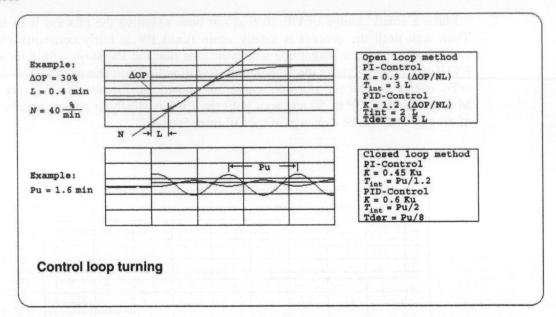

Figure Ex. 7.3
Tuning formula display

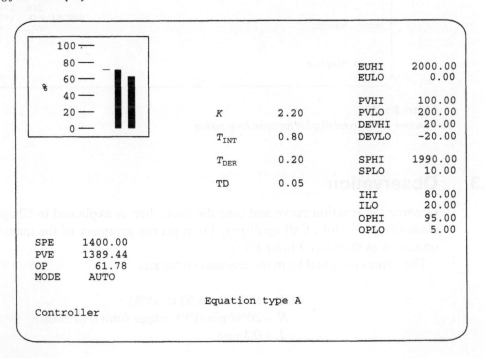

Figure Ex. 7.4
Detail display with tuning constants

E.7.4   Confirmation of proper tuning

Change mode to automatic and make a change of the SP. Observe the process settling down (observe PV). Use the trend display as shown in Figure Ex. 7.5 for observation. Most parameters can be operated from the trend display as well. As a rule of thumb, the settling down should take place with quarter damping (1/4 decay). That means two successive maximum values of a damped oscillation should have a magnitude ratio of 1–4.

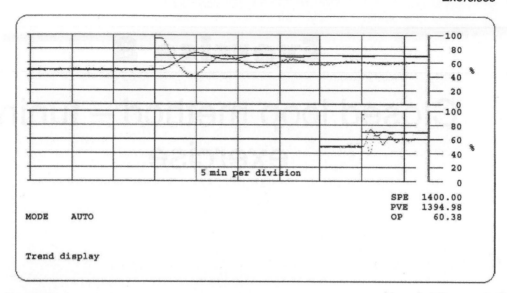

Figure Ex. 7.5
Trend display of tuned loop

E.7.5 Fine tuning

Based on process knowledge (noise on PV etc.), experience and precise knowledge of the control actions of each control mode (P-control, I-control and D-control) fine tuning of the process should be performed now. Only minor variations to the tuning constants should be made.

- *TD-fine tuning*: Observe the noise of the control action (OP). Use your judgment as to how much to increase or decrease TD. The judgment has to be based mainly on the effect the noise in the OP signal has on the manipulated variable (wear and tear of a valve for instance).

  An increase of TD may require an increase of T_2 (derivative time constant) to compensate for the lagging phase shift of the noise filter.

- K-*fine tuning*: Since K is the gain for all control modes (PID), reducing gain reduces the effect of all control actions equally and vice versa. A reduction of K reduces speed of control and increases stability and vice versa.

- T_2-*fine tuning*: After tuning (using the Ziegler–Nichols formulas), the damped oscillation of the process after a change of SP takes place equally around the new SP. Increasing the derivative time constant (T_2) will bend the baseline towards the old value of SP and therefore reduce the overshoot. In addition stability in general will be increased. It has to be noted that the area of error increases as well with increased stability. The reason for this is the slower approach of PV towards the SP.

- T_1-*fine tuning*: In most cases integral tuning should be corrected only in relatively stable loops like flow loops. T_1-fine tuning should be done with close observation using the trend display. Decreasing T_1 increases instability and vice versa.

Exercise 8

Closed loop method – tuning exercise

E.8.1 Objective

This exercise will give some practical experience in the closed loop tuning method. Figure Ex. 8.1 shows the diagram of a typical control loop. The steps for tuning are explained in the Chapter 5 'Tuning of closed loop control systems'.

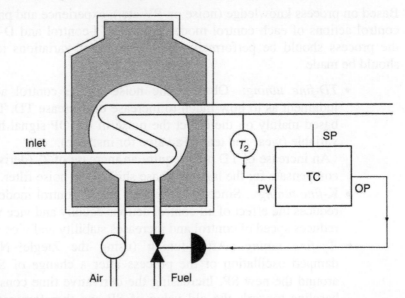

Figure Ex. 8.1
Single loop control

E.8.2 Operation

Call up the training application General *single loop with interactive PID (real form)*.

Then, call up the detail display F3 and change MODE to MANUAL. Set the OP to 50% and wait for the process to settle down. Make a good judgment about noise observed on PV and set the digital noise filter (TD) accordingly. A good value of TD will result in a still noticeable but very small magnitude of noise in the PV term. The noise filter has to be set up before tuning in order to have it included in the results of the tuning exercise. Practically, the noise filter becomes part of the process as far as loop tuning is concerned.

For the closed loop tuning method, we have to change the tuning constants in such a way as to obtain P-control only. Change T_{INT} to 999, T_{DER} to 0 and K to a relatively low value of 0.5. Then, change MODE back to AUTO and make a small change of SPE in order to find out how sensitive the process is in responding. We have to find a value for K so that the closed loop will oscillate with constant magnitude. If the oscillation settles down, K has to be increased and if the oscillation increases in magnitude, K is too large. Having found the value of gain for constant oscillation, we name this particular value of gain K_u (gain of the ultimate frequency). The time period of oscillation we call P_U (period of the ultimate frequency).

To obtain continuous oscillation, as shown in Figure Ex. 8.2, set MODE to AUTO after the loop has settled down. Then change SPE from 1000 to 1100.

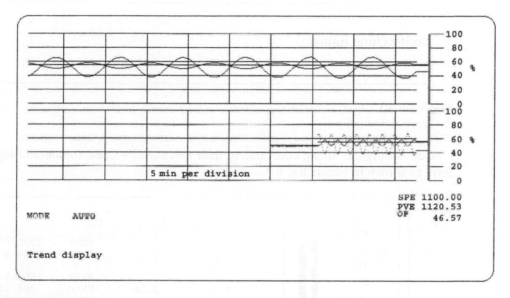

Figure Ex. 8.2
Continuous oscillation trend display for closed loop tuning

E.8.3   Observation

Observe continuous oscillation and tune the controller as explained in Chapter 5 'Tuning of closed loop control'. Call up display F6 to get the summary of the major tuning formulas on screen as shown in Figure Ex. 8.3.

The values obtained from the continuous oscillation trend display are:

$$K_u = 3.5$$
$$P_u = 1.6 \text{ min}$$

Using the formulas shown in display F6, we obtain the following tuning constants for PID-control:

$$K = 2.2$$
$$T_{INT} = 0.8$$
$$T_{DER} = 0.2$$

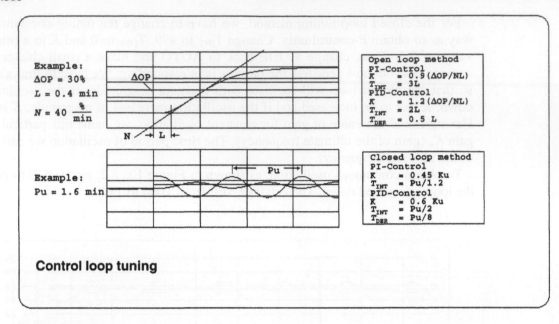

Figure Ex. 8.3
Tuning formula display

After setting the tuning constants to their calculated values, the detail display F3 should look as shown in Figure Ex. 8.4.

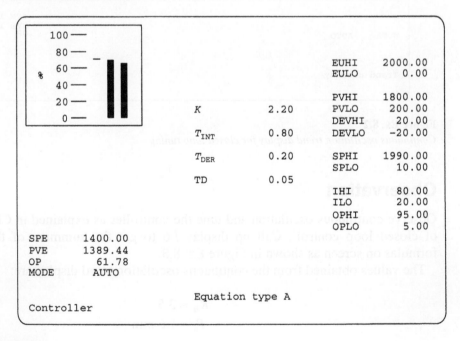

Figure Ex. 8.4
Detail display with tuning constants

E.8.4   Confirmation of good tuning

Wait for the process to settle down and then make a change of SP. Observe the process settling down (observe PV). Use the trend display as shown in Figure Ex. 8.5 for observation. All parameters displayed in the tuning display can be operated from there.

As a rule of thumb, the settling down should take place with quarter damping. This means that two successive maximum values of a damped oscillation should have a magnitude ratio of 1–4.

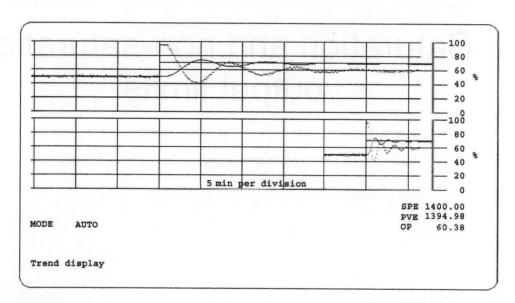

Figure Ex. 8.5
Trend display of tuned loop

E.8.5   Fine tuning

Fine tuning is done in the same way as shown in Exercise 7.

Exercise 9

Saturation and non-saturation output limits

E.9.1   Objective

This exercise will introduce the student to the two major types of OP-limit calculations. The one kind, with OP-limits, allows a saturation value of the OP, based on P-control and D-control, and another kind, does not allow OP values to go into saturation at all.

E.9.2   With saturation of the OP

The controller calculates a VIRTUAL OP value independent of any OP-limit. These may be values far above 100% or far below 0%. Only the real OP, which is the displayed OP value, is limited by OP-limits. The real OP awaits the return of the virtual OP within OP-limits. Then, within the range of the OP-limits, the real OP follows the virtual OP value exactly. Controllers driving field OP normally use this kind of OP-limit handling.

E.9.3   With non-saturation OP-limit-calculation

Only real OP values are used for the calculation. If a single calculation results in an OP value beyond OP-limits, the OP value will be set to the value of the OP-limit it would have violated. When the controller calculates the OP value next time (in the next *scan*), the real OP value (OP = OP-limit) is used. The previous calculation, beyond OP-limits, has been totally forgotten.

E.9.4   Virtual OP limit operation

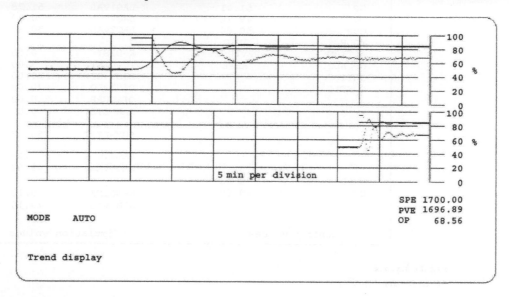

Figure Ex. 9.1
Virtual OP-limit calculation

E.9.5   Operation

Call up the training application 'General single loop with interactive PID (real form)'. In order to operate with virtual OP limit calculation, call display F8 and make sure that the status variable OPCALC is set to VIRTUAL. Then make large changes of SPE (from 1000 to 1700) in order to observe the OP reach saturation. Figure Ex. 9.1 shows how the derivative action goes into saturation.

E.9.6   Real OP limit operation

Without saturation of the OP, there is no difference made between calculated and real OP values. If a calculation of the OP exceeds the OP-limits, no difference is made between calculated and real output. If the result of an OP calculation would be beyond an OP-limit, the real OP value is set to the OP-limit value. No virtual OP value is memorized for the next scan's calculation. Generally, computer-resident PID algorithms, not driving field values, use this kind of OP-limit handling. Before using this kind of OP-limit calculation for this exercise, call up display F8 and change the variable OPCALC to REAL (see Figure Ex. 9.2).

Figure Ex. 9.3 shows how the same SP change causes a different derivative action (without saturation of the OP) as shown in Figure Ex. 9.1 (with OP saturation).

The trend display shown in Figure Ex. 9.3 is difficult to understand. The most important thing to observe in this display is a little single dot at 95% at the same time when SPE changed from 1000 to 1700 in a step function. The single isolated dot represents a single OP calculation limited by OPHI. Already the next calculation of OP includes a decrement of OP so large, that the OP limit OPLO inhibits OP to be below 5% of range.

Think about the differences and think about possible applications for the two kinds of OP limit calculations. *Remember the use of EQ-Type B or C.*

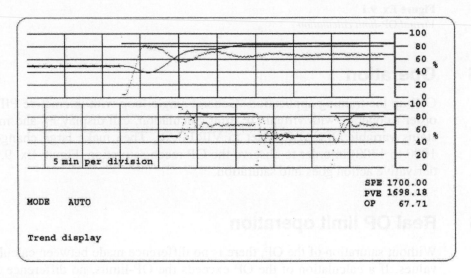

```
OPVIRT          51.41        LAG1TC           0.40
OP              51.41        LAG1VAL         51.78
PV              49.10        LAG2TC           0.40
CSP             50.00        LAG2VAL         50.37
MODE            AUTO
ACTION          REVERSE      K-DIST           0.00
OPCALC.         REAL         TC-DIST          3.00
INIT                         DIST-HI         20.00
CONFIG                       DIST-LO        -20.00
                             DISTURB          0.00
                             RAW-DIST         0.00

                             SUM-DIST        50.37

                             LAG3TC           0.40
                             LAG3VAL         49.62

CVPD           -0.46
CVI            -0.00         K-NOISE          0.10
                             SIM-VAL         49.15

       Control values                Simulation values
```

Figure Ex. 9.2
Auxiliary display (F8)

```
5 min per division

                                    SPE  1700.00
                                    PVE  1698.18
MODE    AUTO                        OP     67.71


Trend display
```

Figure Ex. 9.3
Real OP-limit calculation

Exercise 10

Ideal derivative action – ideal PID

E.10.1 Objective

This exercise will introduce the non-interactive form of the PID algorithm (*Ideal PID*). The ideal PID algorithm makes use of a mathematically true derivative calculation (sT) Figure Ex. 10.1. Generally, ideal PID control is combined with non-saturation OP-limit handling. Ideal PID is used for high level control concepts only (e.g. PID-X in Honeywell equipment). The algorithms reside mainly in supervising computers (as opposed to PLCs, loop controllers or the RTUs of a DCS system). The same SP change can cause a very different (and very confusing) looking derivative control action for either the ideal or real PID controller. Compare the derivative control of a non-interactive PID controller with real OP-limit calculation (Figure Ex. 10.2), to that of an interactive PID controller with real OP-limit calculation (Figure Ex. 10.3). Detailed knowledge about the practical application is required to make optimum use of this kind of control at the right place. Note that if you are in doubt as to which algorithm to choose, select real PID control.

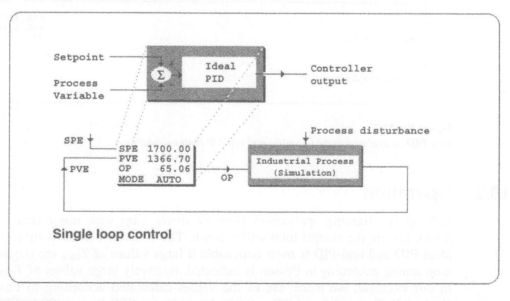

Figure Ex. 10.1
Closed loop control block diagram (Ideal PID)

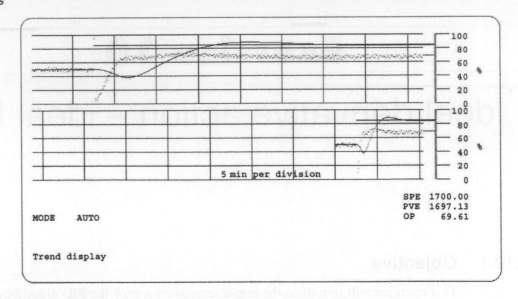

Figure Ex. 10.2
IDEAL PID control with strong OP-*noise and REAL* OP *limit calculation*

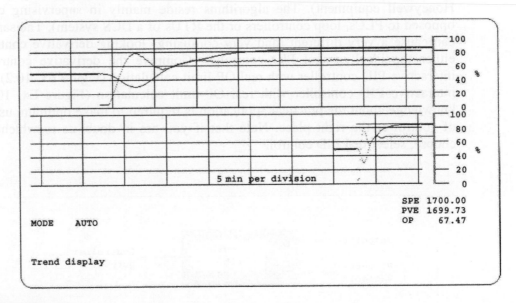

Figure Ex. 10.3
Real PID control with some OP-*noise and real* OP *limit calculation*

E.10.2   Operation

Call up the training application *General single loop with non-interactive PID (ideal form)*. Ensure the control loop settles down. The difference in derivative control between ideal-PID and real-PID is most noticeable if large values of T_{DER} are required. Whenever loop tuning according to Pessen is indicated, relatively large values of T_{DER} will be used. In our exercise, we make use of the values calculated according to Pessen ($K = 1.1$, $T_{INT} = 0.8$, $T_{DER} = 0.5$ and TD = 0.05). Now change SPE from 1000 to 1700 and observe the OP. Figure Ex. 10.2 shows the result.

E.10.3 Observation

The considerable level of noise created by the derivative action of the non-interactive PID (Ideal PID) control should be noted.

E.10.4 Operation

Call up the training application *General single loop with interactive PID (real form)*. Call up display F8 and change the variable OPCALC to REAL. Then, perform *identically* the same operation as before. Figure Ex. 10.3 shows the results.

E.10.5 Conclusion

It can clearly be seen that the ideal form of the PID algorithm (non-interactive PID) creates significantly more noise than the real form. Ideal PID control leads to unacceptable high wear and tear of physical equipment. This makes ideal PID unsuitable for field interaction (non-interactive). Essentially this means that ideal PID control is unsuitable to operate with real world processes or field values (valves, etc.).

E.10.6 Operation

Again, call up the training application *General single loop with non-interactive PID (ideal form)*. Set the tuning constants to the same values as before ($K = 1.1$, $T_{INT} = 0.8$ $T_{DER} = 0.5$ and TD = 0.05). Then, change SPE from 1000 to 1020 and observe the OP. No changes of the tuning constants should be made during the whole exercise.

E.10.7 Observation

A relatively small step function of SPE from 1000 to 1020 results in a derivative action, reaching close to the OPHI limit. The derivative action, as a result of the step function of SPE, takes place for one scan time only (needle pulse). It is visible in Figure Ex. 10.4 as a single dot of OP close to the value of the OPHI limit, 95% of range. The remaining OP

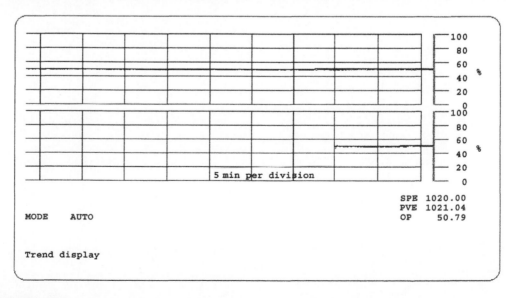

Figure Ex. 10.4
Ideal PID control action

action is based on PID-control as a result of PVE. It should be noted that this occurs with step functions of SPE or PVE only. In every other case, PID control is performed as a continuous function.

E.10.8   Conclusion

Step functions of SPE or PVE result in a needle pulse of OP. These pulses are not received well by field equipment, driven by the output. For example, no valve should experience a single hit by a strong but short pulse of the OP. There is no time for the valve to move and therefore there is no effect on the process. The only result is that the control valve will wear out quickly, when put under such unnecessary high mechanical stress. One important conclusion that comes from this is that you should: consider the use of Equation Type B or C, where the operator is not able to influence PD-control via step changes of the SPE. The PVE hardly makes step changes.

Discuss the differences and think about possible applications for ideal and real PID algorithms. *Remember the use of EQ-Type B and C*. Special attention has to be given to noise. This exercise makes it very clear why the real PID algorithm is preferred over the ideal PID algorithm, if used as the ultimate secondary controller (or field controller).

Exercise 11

Cascade control

E.11.1 Objective

The objective of this exercise is not to provide a guided tour through cascade control, but to introduce the student into the effects of PV tracking, initialization and Mode changes in cascade control. The other aspects of cascade control can easily be explored with the capabilities of this software package. But it is up to the student to experiment.

E.11.2 Operation

Call up the training application *Tank level control* (see Figure Ex. 11.1). At this stage no PV tracking or initialization is active. After the process has settled down, change MODE2 to MANUAL and OP2 to 30% of range. The objective is to continue cascade control when PVE1 has reached approximately a value of 70 (35% of range). As we can see in Figure Ex. 11.2 (see left side of the screen), the change of MODE2 and OP2 causes most variables within the cascade control system to drift uncontrolled (OP1, SPE1, SPE2). When PVE1 has reached approximately 70, change SPE1 to 70 and then MODE2 back to CASC.

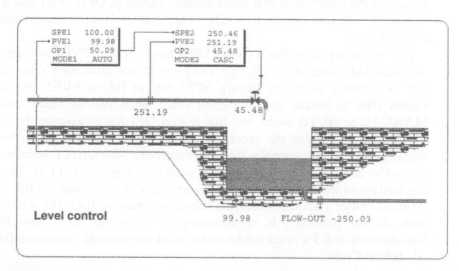

Figure Ex. 11.1
Level control block diagram (F2)

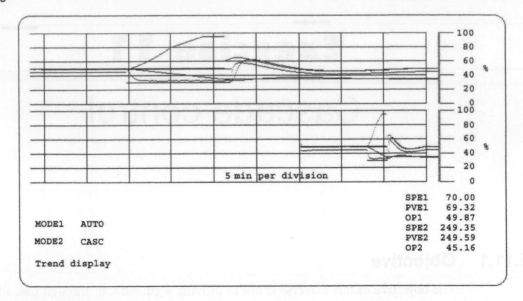

Figure Ex. 11.2
Trend without PV tracking and initialization

E.11.3　Observation

When we change from CASC control to MANUAL (MODE2 from CASC to MANUAL) some variables are unpredictable as discussed earlier. When changing back to CASC control, these unpredictable variables cause a bump within our control system, when they have to change back to real and defined values.

E.11.4　Conclusion

Based on the observation that unpredictable values of OP1, SPE1 and SPE2 cause a bump in control when we change to CASC control, the following requirements have to be met to avoid this.

The secondary controller's (flow) SPE2 has to follow PVE2 as long as the controller is in MANUAL mode (PV tracking). The same requirement for PV tracking has to be made for the primary controller (level). SPE1 has to follow PVE1. In addition, we cannot permit OP1 to assume any unwanted value when the secondary controller is either in MANUAL or AUTO mode. For this reason OP1 has to assume the same value in % as SPE2 has, as long as the secondary controller is not in CASC mode. This is called initialization. In our example, the flow controller (secondary controller) is the initializing controller and has to be configured as such (see Figure Ex. 11.3). The level controller is the initialized controller. If the level controller is initialized by the flow controller, the level controller's MODE indication shows for example I-AUTO, which means 'initialized with AUTO pending'. It is necessary to have the flow controller configured for initialization and PV tracking in order to obtain smooth transition from MANUAL (or AUTO) to CASC.

```
      OPVIRT1          49.96          K-NOISE          0.40
      OP1              49.96
      PV               49.92          VALUETC          0.10
      CSP1             50.00          K-FL             1.10
      MODE1            AUTO           FL-SIM          49.57
      ACTION1          REVERSE
      OPCALC1          VIRT           AVG-OUT        -50.00
      INIT1                           K-DIST           0.00
      CONFIG1          TRACKING       TC-DIST          0.30
                                      DIST-HI         20.00
      OPVIRT2          45.33          DIST-LO        -20.00
      OP2              45.33          DISTURB          0.00
      PV               50.05          RAW-DIST         0.00
      MODE2            CASC
      ACTION2          REVERSE        FL-OUT%        -49.66
      OPCALC2          VIRT
      INIT2                           TANK-TC          2.50
      CONFIG2          I & TR         LVL-SIM         49.92


         Control values                  Simulation values
```

Figure Ex. 11.3
Auxiliary display to configure CONFIG1 and CONFIG2 (F8)

E.11.5 Operation

In order to satisfy the above requirements, go to display F8 and change CONFIG1 to TRACKING and CONFIG2 to I and TR (initialization and tracking). Change SPE1 to 100 (50% of range), MODE2 to CASC, MODE1 to AUTO and wait until the process has settled down. Change MODE2 to MANUAL. Then change OP2 to 30% of range. When SPE1 has reached approximately 70, change MODE2 back to CASC.

E.11.6 Conclusion

Based on the observation that unpredictable values of OP1, SPE1 and SPE2 cause a bump in control when we change to CASC control, the previous requirements have to be met to avoid this problem.

E.11.7 Observation

As seen in Figure Ex. 11.4, the values of OP1, SPE1 and SPE2 assume meaningful values at all times when MODE2 is MANUAL. The whole cascade control system is ready to control in CASC mode. No bump occurs at time of transition from MANUAL to CASC (MODE2).

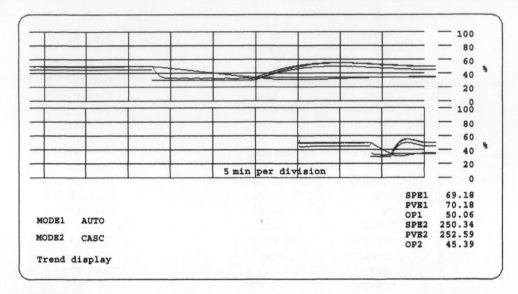

Figure Ex. 11.4
Trend display with PV tracking and initialization

E.11.8   Conclusion

We come to the conclusion, that it is advisable to make use of PV-tracking and initialization whenever possible. Special care is necessary when deciding on the use of PV-tracking in a primary controller. The incorrect use of PV-tracking may cause some operational problems. This is explored in more detail in Exercise 13.

Exercise 12

Cascade control with one primary and two secondaries

E.12.1   Objective

The objective of this exercise is to provide some experience in *Cascade control with multiple secondary controllers*. Although various aspects are examined in some detail in this section, the software allows the student to explore other areas in more exhaustive detail.

E.12.2   Training application

The application used in this exercise deals with cascade control, using two secondary controllers and one primary controller. In order to drive the setpoints of both secondary controllers independently from each other by the same primary controller, two independent OP calculations take place in the primary controller. These independent OP calculations result in two output values OPA and OPB of the primary controller and independent modes MODEA and MODEB (see Figure Ex. 12.1).

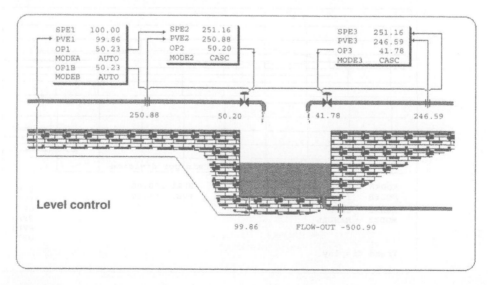

Figure Ex. 12.1
Block diagram

Generally, the amount of data for multiple outputs is too much for one display. As the primary controller in our training application has only two outputs, both are always displayed together on the same display. Viewing both outputs, their modes, alarms and initialization conditions on one screen is more useful in this (teaching) environment.

E.12.3   Operation

Call up the training application *Tank level control with two inlets*. Call the block diagram F2 and study the control concept without changing any values. Then, select trend display F4.

E.12.4   Observation

The training application is in a fully operational and stable automatic control status and correctly tuned when first retrieved. PVE2 and PVE3, the process variables of the two flow controllers, are approximately equal at the beginning although OP2 and OP3 are not equal. OP2 is approximately 50% and OP3 is approximately 40%. Various reasons for this could include (minor) differences in the supply pressures, valve sizes and pipe diameters.

E.12.5   Operation

Change MODE3 to MANUAL and then OP3 to 30% of range.

E.12.6   Observation

We can see that MODEB has changed to I-AUTO and MODEA is still in AUTO. This is an indication that OPB is initialized because MODE3 is in MANUAL. We can actually observe initialization of OPB taking place. OPA is continuing to drive SPE2 and automatic level control still takes place by controlling the flow PVE2 only. As PVE3 has been reduced by manually changing OP3 to 30%, the level starts to drop. The level controller compensates for this by raising OP2 via the flow controller (see Figure Ex. 12.2).

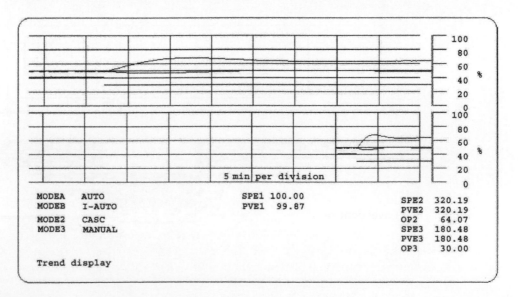

Figure Ex. 12.2
Manual OP3 from 40 to 30%

Figures Ex. 12.2, 12.3, 12.4 and 12.5 show four trend pens only. These are trends for SPE1, PVE1, OP2 and OP3. The trend pens for SPE2, PVE2, SPE3 and PVE3 have been suspended using the commands TREND2, TREND3, TREND5 and TREND6. If you require all the trend pens on screen simultaneously when you execute this exercise, then don't use the TRENDx command.

E.12.7 Operation

Change OP3 even further, from 30 to 10% of range.

E.12.8 Observation

The level controller makes an attempt to compensate for the reduction of flow (PVE3) by increasing OP2 via OPA. This attempt is unsuccessful, as OP2 and OPA violate their integral high limits (IHI = 80%). The control action based on one flow controller only is not strong enough and the tank level drops slowly. In addition, we see an alarm DEV-LO, which tells us that PVE2 deviates too much from SPE2. This alarm has been raised as no integral action can take place beyond the integral limits. Therefore, no control without offset is possible (see Figure Ex. 12.3).

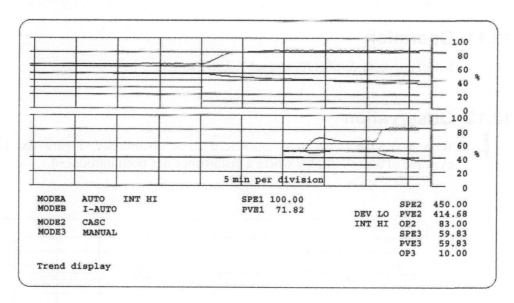

Figure Ex. 12.3
Manual OP3 from 30 to 10%

E.12.9 Operation

Change MODE3 to CASC.

E.12.10 Observation

We can now see that both flow controllers are working together again. As OPA and OPB are calculated independently. OPA is observing the integral limit IHI as long as OPB increases OP3 to a level as to cause OPA to return to within its integral limits. From then on, we can observe that both OP2 and OP3 behave dynamically in an identical manner, as long as no output violates any output limit (including the integral limits) (see Figure Ex. 12.4).

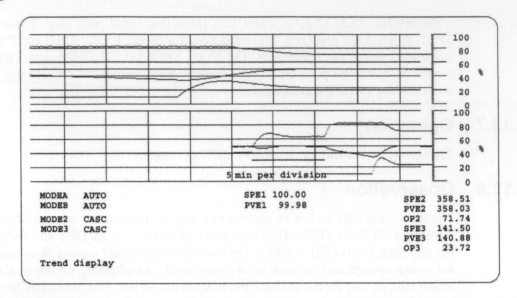

Figure Ex. 12.4
Return of MODE3 to CASC

E.12.11 Operation

Select auxiliary display F8 and set K-DIST to 1.

E.12.12 Observation

With K-DIST = 1, we have introduced some disturbance. Figure Ex. 12.5 shows an example of a trend display where a disturbance has been introduced.

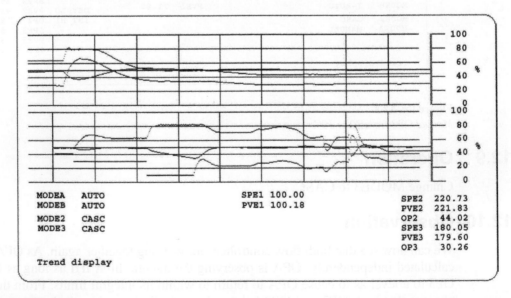

Figure Ex. 12.5
Control with disturbance

E.12.13 Conclusion

The most important fact that we can glean from this exercise is that automatic control takes place as long as at least one secondary controller was in cascade mode (CASC). It can clearly be seen that both outputs OPA and OPB are calculated totally independent of each other.

Exercise 13

Combined feedback and feedforward control

E.13.1   Objective

This exercise is provided to explore combined feedback and feedforward control. The intention of this section is to introduce this control strategy. Although it is not possible to go into all aspects of this topic in this exercise, the control of the training applications have not been simplified. The control examples provided are 'feedheater control' (Figure Ex. 13.1) and 'boiler level control' (Figure Ex. 13.2).

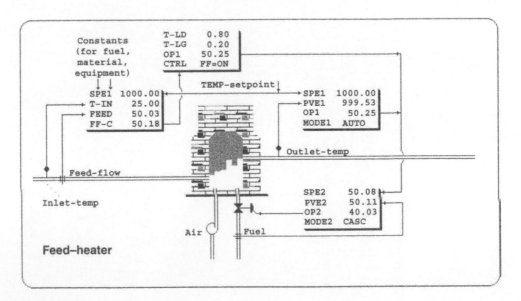

Figure Ex. 13.1
Combined feedback and feedforward control of a feedheater

A few points to remember

The main control is feedforward control, whenever all major disturbances are used to calculate feedforward control. In most cases, feedback control serves as a long term correction. In essence this means that it should act against the slow drift of the PV from the setpoint which feedforward cannot be expected to correct. Therefore, tuning should be done in the following order. Firstly, the flow controller which is common for feedback and feedforward control has to be tuned. Secondly, feedforward control and finally feedback control has to be tuned. The tuning of feedback control should aim towards minimum feedback control action, just enough to eliminate the process drift. Anything more adds to the wear and tear of equipment without significantly improving control results.

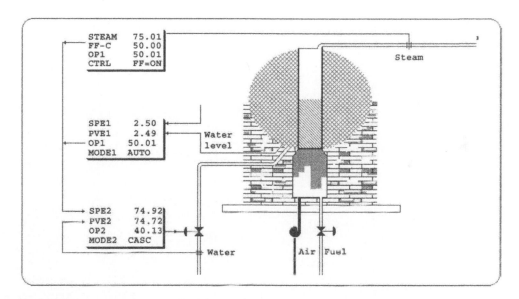

Figure Ex. 13.2
Combined feedback and feedforward control of a boiler

E.13.2   Common pitfalls

Combined feedforward and feedback control often makes use of one value in different places. If the operator or engineer is not aware of it, changes of those values may be made with one objective in mind and with a hidden surprise caused by an unknown link.

Feedheater control example:
SPE1 is used for feedback control within the temperature controller as setpoint and in addition the same SPE1 is used in feedforward control to calculate the temperature difference (SPE1) – (T-IN). The calculation is target temperature minus inlet temperature and is proportional to the amount of fuel required to heat from T-IN to SPE1. This calculation of temperature difference will be corrupted totally, if SPE1 changes unexpectedly because PV-tracking has been configured for the temperature controller.

E.13.3   Special effects with boiler level control

The level controller must be tuned for slow control reactions only. Sudden load changes of steam make the water level in boilers appear to change in an opposite direction to that expected. The sudden increase in steam flow decreases boiler pressure. As a result of decreased pressure, the water is boiling more vigorously, which causes the water level to rise. This happens only for short periods of time as a transitional effect, until normal pressure has been restored. If the level controller would react upon those transitional level changes, the control action would go in the wrong direction. A further explanation of this effect is that a sudden increase of steam load causes a sudden increase of evaporation and the bubbles of steam in the water increase. This causes an increase in the water level and a decrease of the water volume in the boiler. Due to this effect, fast flow control should be based on feedforward control (massflow steam = massflow water) and long term level control should be based on feedback control only. This is done in effect, if you follow the above advice.

Exercise 14

Deadtime compensation in feedback control

E.14.1 Objectives

This exercise will give some practical experience in the closed loop control strategy necessary if a long deadtime is part of the process to be controlled. Figure Ex. 14.1 shows the block diagram of a control loop with added process simulation as a means of overcoming process deadtime.

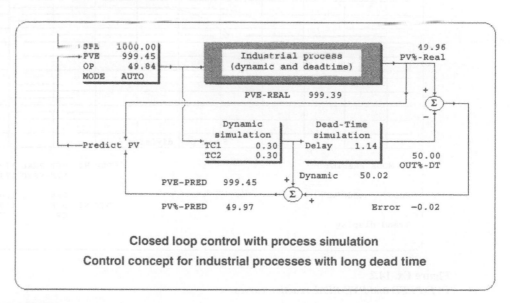

Figure Ex. 14.1
Block diagram

Note concerning terminology

Within this exercise, we have a conflict of terminology. As we have no physical industrial plant to control, we have to simulate the behavior of such an industrial plant. If one of the tools of controlling an industrial plant requires a process simulation, we would not know what simulation is meant if we do not make a clear distinction between the process simulation acting as a stand-in on behalf of the real process, and the simulation acting as a tool of control.

Therefore, in the description of this exercise, the term simulation is never used if referring to the real process; the terms for the real industrial process will be used here instead. In this exercise, the term simulation is used only for a process simulation which exists in addition to the industrial process and would exist in a real industrial plant for control purposes as well.

E.14.2 Operation

Select the training application *Deadtime compensation in single loop control*. The tuning constants for the process simulation are in detail display F5. They are K-SIM, BIAS-SIM, TC1, TC2 and DELAY. The values have been initialized to the correct values required to match the real process as close as possible.

Then, select the trend display F4 and change MODE to MANUAL. Set the OP to 50% and wait for the process to settle down. Make a change of OP from 50 to 70% of range (see Figure Ex. 14.2).

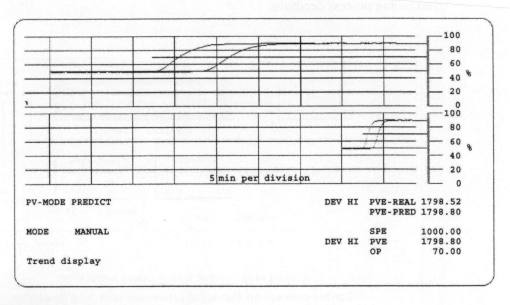

Figure Ex. 14.2
PVE-PRED and PVE-REAL reaction curves

E.14.3 Observation

Observe the different reaction curves of the process variable coming from the real process (PVE-REAL) and simulated process variable (PVE-PRED). It can be seen that the reaction of the predicted process variable is similar, but if compared to the real process variable there is no deadtime.

E.14.4 Operation

In order to explore the purpose and impact of the simulation tuning constants, do the following for each tuning constant (for K-SIM, BIAS-SIM, TC1, TC2 and DELAY):

- Make sure, the controller is in MANUAL mode
- Change the tuning constant
- Make a step change to the OP
- Carefully observe the reaction curves of PVE-REAL and PVE-PRED
- Return the tuning constant to its correct value
- Repeat the above operations with different values for each tuning constant. Avoid experimenting with more than one tuning constant at any one time in order not to confuse the results.

E.14.5 Observation

The reaction curve of PVE-PRED changes as follows when the tuning constants are changed:

- The change of K-SIM changes the magnitude
- The change of BIAS-SIM raises or lowers the base line
- The change of TC1 and TC2 change the dynamic of the simulation (form of reaction curve)
- The change of DELAY causes significant deformations of the reaction curve of PVE-PRED. The characteristic of these deformations indicate whether the deadtime DELAY is too long or too short. Compare the reaction curve obtained with the tuning display F7 (see Figure Ex. 14.3).

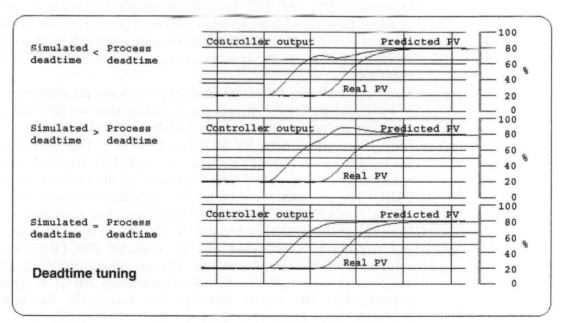

Figure Ex. 14.3
Deadtime tuning

E.14.6 Conclusion

K-SIM, BIAS-SIM, TC1 and TC2 shape the simulated reaction curve of PVE-PRED. DELAY serves to match the simulated deadtime with the process deadtime.

E.14.7 Operation of simulation tuning

Select detail display F5 and make sure that the controller is in MANUAL mode. K-SIM, BIAS-SIM, TC1, TC2 and DELAY are the tuning constants for the process simulation. They have to be set to match the real process as close as possible.

Note: The real process can never be matched by simulation with mathematical precision. Therefore, we have to take great care in order to match the simulation as close as possible to the real process. In order to have a realistic training situation, it is not possible to obtain an ideal match within this exercise, but you can come close to it. As a preparation for this tuning exercise, change K-SIM, BIAS-SIM, TC1, TC2 and DELAY to values between 0 and 0.1. This makes sure that the simulation is totally out of tune and tuning is required.

Tuning should be done in the following sequence:

- Use the control loop without simulation and try to tune it using the 'Closed Loop Method' according to either Ziegler and Nichols or Pessen. In order to have a classical closed loop without any simulation, change the status variable PV-MODE to REAL. Then, the simulation is still calculated and displayed but is not used for control purposes. This step provides us with a general idea of safe tuning constants (K, T_{INT} and T_{DER}) of the closed loop.

- Switch the controller mode to MANUAL. Make step changes to the OP and experiment with the tuning constants K-SIM and BIAS-SIM until the correct values have been found. The values for K-SIM and BIAS-SIM are correct if the final value of PVE-REAL and PVE-PRED approach the same value without bias towards the end of both reaction curves.

- Manipulate TC1 and TC2 in order to match as closely as possible the dynamic behavior of both reaction curves. The observation of the reaction curve of PVE-PRED is very difficult because good judgment is necessary to separate the effect of the dynamic changes from the effect of non-matching deadtimes.

- Match the deadtime of the real industrial process and the simulation. Make use of the tuning display F7. Repeat the preceding step and this step as often as necessary to obtain an optimum match of dynamic and deadtime.

- Change the status variable PV-MODE to PREDICT. This makes the controller use the predicted variable PVE-PRED instead of PVE-REAL. Obtain optimum tuning constants for closed loop control using the predicted variable PVE-PRED. You may either use the method according to Ziegler and Nichols or Pessen. Detail display F3 contains the necessary tuning constants.

- Even though the stability problem of the closed loop is dramatically reduced, it still migrates to some extent into the predicted PVE-PRED via the error calculation of the control strategy. The magnitude of this impact can be subjected to a complex mathematical evaluation but is in practice quite unpredictable. The reason therefore lies mainly in the unaccountable differences between process and simulation.

  Hence, in order to make sure that the problems discussed above don't cause unpleasant and unexpected surprises, a long time of observation of any unexpected stability problems is essential. As a guide, the time of observation should be at least discussed above 20 times the deadtime.

- Based on the observation made in the preceding step, the tuning constant K has to be lowered to a point where smooth control is assured.

E.14.8 Conclusion

The basic idea of deadtime compensation is straightforward. Nevertheless, it is important to realize that it is a strategy of intermeshed loops. The control loop using the predicted PVE-PRED is intermeshed with remnants of the real PVE-REAL via the error calculation. In addition, one has to realize that the reaction of control actions is predicted only. No prediction whatsoever can be made about the impact of disturbances. Derivative control is most effective in counteracting integral and lagging behavior, but has its limitations if applied to deadtime problems. Therefore, one should refrain from using PID-control and accept that PI-control is the preferred solution in these cases.

Exercise 15
Static value alarm

E.15.1 Objective

The objective of this exercise is to demonstrate the use of statistics in process control. In some cases of pressure measurement, a pressure meter ceases to provide correctly updated values. The last correct value obtained is provided continuously as a static value which may still be within its correct range and not be recognized by either operator or automatic alarm mechanisms as being an incorrect value. This exercise makes use of a continuous (running) standard deviation calculation. An alarm will be raised whenever no changes of the measured value take place. The concept of detection of a false value is based on the premise that the standard deviation value is too low (see Figure Ex. 15.1).

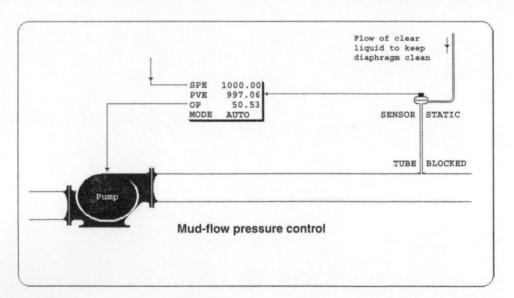

Figure Ex. 15.1
Block diagram

E.15.2 Operation

Select the training application *Pressure control with static sensor alarm*. In order to observe the operation of the static sensor alarm, select display F2 and simulate a blockage of the sensor pipe. To simulate a blocked or free sensor pipe, manipulate the status variable TUBE. TUBE can be set to either status-FREE or BLOCKED (see Figure Ex. 15.1).

E.15.3 Observation

When the sensor pipe is blocked, the status variable SENSOR displays the alarm STATIC. Immediately after removing the blockage, the alarm disappears and accurate readings of PVE are available.

E.15.4 Conclusion

This example demonstrates that it is worthwhile to explore the use of statistical means in order to improve some aspects of process control.

Index

Other titles in the series

Practical Cleanrooms: Technologies and Facilities (David Conway)

Practical Data Acquisition for Instrumentation and Control Systems (John Park, Steve Mackay)

Practical Data Communications for Instrumentation and Control (Steve Mackay, Edwin Wright, John Park)

Practical Digital Signal Processing for Engineers and Technicians (Edmund Lai)

Practical Electrical Network Automation and Communication Systems (Cobus Strauss)

Practical Embedded Controllers (John Park)

Practical Fiber Optics (David Bailey, Edwin Wright)

Practical Industrial Data Networks: Design, Installation and Troubleshooting (Steve Mackay, Edwin Wright, John Park, Deon Reynders)

Practical Industrial Safety, Risk Assessment and Shutdown Systems for Instrumentation and Control (Dave Macdonald)

Practical Modern SCADA Protocols: DNP3, 60870.5 and Related Systems (Gordon Clarke, Deon Reynders)

Practical Radio Engineering and Telemetry for Industry (David Bailey)

Practical SCADA for Industry (David Bailey, Edwin Wright)

Practical TCP/IP and Ethernet Networking (Deon Reynders, Edwin Wright)

Practical Variable Speed Drives and Power Electronics (Malcolm Barnes)

Practical Centrifugal Pumps (Paresh Girdhar and Octo Moniz)

Practical Electrical Equipment and Installations in Hazardous Areas (Geoffrey Bottrill and G. Vijayaraghavan)

Practical E-Manufacturing and Supply Chain Management (Gerhard Greef and Ranjan Ghoshal)

Practical Grounding, Bonding, Shielding and Surge Protection (G. Vijayaraghavan, Mark Brown and Malcolm Barnes)

Practical Hazops, Trips and Alarms (David Macdonald)

Practical Industrial Data Communications: Best Practice Techniques (Deon Reynders, Steve Mackay and Edwin Wright)

Practical Machinery Safety (David Macdonald)

Practical Machinery Vibration Analysis and Predictive Maintenance (Cornelius Scheffer and Paresh Girdhar)

Practical Power Distribution for Industry (Jan de Kock and Cobus Strauss)

Practical Hydraulics (Ravi Doddannavar, Andries Barnard)

Practical Power Systems Protection (Les Hewitson, Mark Brown and Ben. Ramesh)

Practical Telecommunications and Wireless Communications (Edwin Wright and Deon Reynders)

Practical Troubleshooting Electrical Equipment (Mark Brown, Jawahar Rawtani and Dinesh Patil)

Practical Batch Process Management (Mike Barker and Jawahar Rawtani)

Printed and bound by CPI Group (UK) Ltd, Croydon, CR0 4YY
03/10/2024
01040338-0019